Geology of the

The district described in this memoir lies toward the western end of the Midland Valley; it includes the greater part of Glasgow and the towns of Clydebank, Paisley, Johnstone and Barrhead, together forming one of the most densely populated areas in Scotland. By contrast, the Kilpatrick Hills and Campsie Fells, which form a belt 6 km wide running across the district, are very sparsely populated and mostly given over to hill farming and forestry. North of the hills, the land gradually falls away to the rich farming land of the Endrick valley.

The oldest rocks that crop out in the district were deposited during the early Devonian by a westward-flowing river system in an intermontane basin which covered a large part of the Midland Valley. After a period of widespread earth movements accompanied by uplift and erosion during middle Devonian times, the palaeoslope was reversed and sandstones of late Devonian age were laid down in a depositional basin dominated by eastward-flowing rivers. Aeolian sandstones are interbedded in the upper part reflecting the arid climate of that time. A return to fully fluvial sedimentation in the early Carboniferous, with sandstones being deposited on the coastal floodplain of a major river system, implies a change in climate at about the Devonian–Carboniferous boundary. Marginal marine conditions became established for a time in the late Tournaisian, prior to the recommencement of fluvial sedimentation as renewed uplift took place in the Highlands. In mid-Dinantian times, sedimentation was interrupted by a major episode of volcanicity during which a thick sequence of subaerial, mainly alkaline basaltic lavas was erupted, typical of that which occurs in continental rift systems throughout the world. After the cessation of volcanicity in the late mid-Dinantian, cyclothemic sequences accumulated in fluviodeltaic, lacustrine and fluvial environments, punctuated by frequent short-lived marine transgressions, throughout most of the Carboniferous. During the deposition of the Limestone Coal Formation and the Coal Measures, lower delta-plain and deltaic conditions prevailed allowing frequent and sometimes prolonged colonisation of the delta top by plants which were subsequently preserved as coal. The Carboniferous rocks were then invaded by two groups of igneous intrusions. The earlier group comprises late Carboniferous–early Permian quartz-dolerite sills and dykes which occur in the northern part of the district. The later group is a complex of olivine-dolerite sills of mid-Permian age which only occurs in the southern part of the district.

The final chapter in the geological history of the district was the repeated glaciations during the Quaternary, the effects of which are clearly displayed in the present landscape.

Cover photograph
University of Glasgow on drumlin of Wilderness Till Formation. (D 1978) (Photographer: F I McTaggart)

Glasgow Cathedral and the Necropolis. The Necropolis is built on a crag and tail, the crag being formed by an olivine-dolerite sill (D 4980A).

BRITISH GEOLOGICAL SURVEY

I H S HALL
M A E BROWNE
I H FORSYTH

Geology of the Glasgow district

Memoir for 1:50 000 Geological Sheet 30E (Scotland)

CONTRIBUTORS

M J Arthur
P J Brand
D K Graham
N S Robins
D Stephenson

London: The Stationery Office   1998

© NERC copyright 1997

First published 1997

The grid used on figures is the National Grid taken from the Ordnance Survey maps numbers 57 and 64.
© Crown copyright reserved
Ordnance Survey licence no. GD272191/1997

ISBN 0 11 884534 9

Bibliographical reference
HALL, I H S, BROWNE, M A E, and FORSYTH, I H. 1998. Geology of the Glasgow district. *Memoir of the British Geological Survey*, Sheet 30E (Scotland).

Authors
M A E Browne, BSc, CGeol
British Geological Survey, Edinburgh

I H Forsyth, BSc
I H S Hall, BSc
formerly British Geological Survey, Edinburgh

Contributors
D Stephenson, BSc, PhD
British Geological Survey, Edinburgh

M J Arthur, MSc, DIC
P J Brand, BSc
D K Graham, BA
formerly British Geological Survey, Edinburgh

N S Robins, BSc, MSc, CGeol
British Geological Survey, Wallingford

Other publications of the Survey dealing with this and adjoining districts

BOOKS
British Regional Geology
The Midland Valley of Scotland, 3rd edition, 1985
Memoirs
The economic geology of the Central Coalfield of Scotland, Area I, 1937, 1920, Area V, 1926 and Area VII, 1920
Geology of the Greenock district, 1990
Geology of the Airdrie district, 1996
Geology of the Ben Lomond district, in press
Geology of the Hamilton district, 1997
BGS Report
Lithostratigraphy of the late Devonian and early Carboniferous rocks in the Midland Valley of Scotland. Vol. 18, No. 3, 1986

MAPS
1:625 000
United Kingdom (North Sheet)
    Solid geology, 1979
    Quaternary geology, 1977
    Aeromagnetic anomaly, 1972
    Bouguer anomaly, 1981
    Regional gravity, 1981
1:250 000
Argyll (Sheet 56N 06W)
    Solid geology, 1987
    Aeromagnetic anomaly, 1981,
    Bouguer anomaly, 1979
Clyde (Sheet 55N 06W)
    Solid geology, 1985
    Aeromagnetic anomaly, 1980
    Bouguer anomaly, 1985
Tay-Forth (Sheet 56N 04W)
    Solid geology, 1986
    Aeromagnetic anomaly, 1981
    Bouguer anomaly, 1979
Borders (Sheet 55N 04W)
    Solid geology, 1986
    Aeromagnetic anomaly, 1980
    Bouguer anomaly, 1981
1:50 000
Sheet 30E (Glasgow) Solid, 1994; Drift, 1995
Sheet 22 (Kilmarnock) Solid, 1928; Drift, 1928
Sheet 23 (Hamilton) Solid, 1929; Drift, 1929
Sheet 23W (Hamilton) Drift, 1995; Solid, 1995
Sheet 31W (Airdrie) Solid, 1992; Drift, 1992
Sheet 38W (Ben Lomond) Solid, 1987
Sheet 30W (Greenock) Solid, 1990; Drift, 1989

Printed in the UK for The Stationery Office
J29544 C6 12/97

CONTENTS

One Introduction 1
Two Devonian 3
Lower Devonian 3
Upper Devonian 3
Three Carboniferous — general 6
Classification 6
  Lithostratigraphy of the Lower Carboniferous 6
  Lithostratigraphy of the Upper Carboniferous 6
  Chronostratigraphy and biostratigraphy 8
Four Lower Carboniferous (Dinantian) 9
Inverclyde Group 9
  Kinnesswood Formation 9
  Ballagan Formation 13
  Clyde Sandstone Formation 14
Strathclyde Group 15
  Clyde Plateau Volcanic Formation 15
  Kirkwood Formation 27
  Lawmuir Formation 27
Clackmannan Group (Dinantian part) 31
  Lower Limestone Formation 31
Five Upper Carboniferous (Silesian) 34
Clackmannan Group (Silesian part) 34
  Limestone Coal Formation 34
  Upper Limestone Formation 37
  Passage Formation 39
Coal Measures 40
  Lower Coal Measures 40
  Middle Coal Measures 40
  Upper Coal Measures 42
Six Carboniferous palaeontology 43
Seven Late- and post-Carboniferous intrusive rocks 48
Quartz-dolerite dykes 48
Alkali dolerite sills 48
Alkali dolerite dykes 49
Milngavie sills 49
Eight Structure 50
Regional context and implications 50
Fold structures 51
Faults 53
Nine Quaternary 54
Summary of late Quaternary history 54
Devensian topography 55
Pre-Dimlington Stadial deposits 57
Dimlington Stadial Ice Sheet 63
Windermere Interstadial: deglaciation 64
Loch Lomond Stadial 72
Flandrian 78

Ten Economic geology 83
Resources — solid 83
Resources — drift 84
Resources — groundwater 85
Geological hazards 86
Eleven Geophysical investigations 88
Gravity 88
Magnetic 91
Seismic 91
Geothermal 92
Physical properties 92
Interpretation profile 93
References 96
Appendices
1 List of BGS boreholes in the Glasgow district and adjacent areas that are cited in the text 102
2 1:10 000 maps (Solid) 103
3 1:10 000 maps (Drift) 103
4 List of Geological Survey photographs 104
5 Index of Carboniferous and Devonian taxa 107
6 Distribution and index of Quaternary taxa 108
Index 111

FIGURES

1 Geological map of the Glasgow district x
2 Generalised sequence of Devonian and Carboniferous rocks in the Glasgow district 2
3 Reconstruction of Stratheden Group palaeogeography in central Scotland 5
4 Correlation of the Inverclyde Group in the Glasgow district 10
5 Molar percent $CaCO_3$ in carbonates 12
6 Stratigraphical divisions of the Clyde Plateau Volcanic Formation in different lava blocks 20
  a. Campsie Fells
  b. Kilpatrick Hills
  c. Beith–Barrhead Hills
7 Composite sections showing the correlation of the Clyde Plateau Volcanic Formation in the Kilpatrick Hills and with adjacent areas 25
8 Compositional range of lavas within the Glasgow district 26
9 Comparative vertical sections of the Lawmuir Formation 28
10 Comparative borehole sections to show variation in thickness in the Lower Limestone Formation 32

11 Vertical sections of the Limestone Coal Formation 35
   a. Comparative sections of the lower part showing eastward thickening
   b. Generalised section of the upper part
12 Comparative borehole sections in the Upper Limestone Formation 38
13 Generalised vertical sections of the Lower Coal Measures 41
14 Vertical section of the Middle Coal Measures up to the Glasgow Upper Coal 42
15 Location of the mid-Dinantian unconformity 51
16 Generalised structure of the Glasgow district showing the principal faults and folds 52
17 Location plan for the Quaternary of the Glasgow district and adjacent areas 55
18 Ice-flow and other glacial features in the Glasgow district 59
19 Sections of the Cadder and Wilderness Till formations near Bishopbriggs 60
20 Sections of the Quaternary deposits in the Erskine Bridge, Linwood and Bridgeton boreholes 62
21 Deglaciation at the end of the Dimlington Stadial 65
22 Glaciation and deglaciation of the Endrick valley area during the Loch Lomond Stadial 67
23 Horizontal section, Shieldhall, Glasgow 68
24 Generalised section in Windmillcroft Dock 68
25 Section at Stobcross railway cutting 69
26 Generalised section of the Quaternary deposits at Inchinnan 71
27 Sections of the Quaternary deposits in the Mains of Kilmaronock, Gartness and Killearn boreholes 73
28 Generalised distribution of Flandrian estuarine flats in the lower Clyde estuary 80
29 Geophysical maps of the Glasgow district and adjacent areas 88
   a. Bouguer gravity anomaly map
   b. Aeromagnetic total field anomaly map
30 Main geophysical features of the Glasgow district and adjacent areas 89
31 Gravity and magnetic profile across the Glasgow district and surrounding areas 94

TABLES

1 Classification of the Carboniferous in Scotland 7
2 Determinations of various metals in acid soluble (carbonate) fraction of carbonate-rich rocks 11
3 Classification of basic igneous rocks 18
4 Stratigraphy of the Clyde Plateau Volcanic Formation in the western Campsie Fells 19
5 Stratigraphy of the Clyde Plateau Volcanic Formation in the Kilpatrick Hills 21
6 Divisions of the Beith–Barrhead lava succession 23
7 Lithostratigraphy of the Quaternary formations 58
8 Stability related to stoop-and-room workings 86
9 Representative physical properties for saturated rocks at ground level 90

PLATES

Cover photograph   University of Glasgow on drumlin of Wilderness Till Formation
Frontispiece   Glasgow Cathedral and the Necropolis
1 Upper Devonian aeolian sandstone, Finnich Glen, near Killearn 4
2 Partly silicified mature cornstone in the Kinnesswood Formation, Roughting Burn, near Dumbarton 9
3 Ballagan Formation overlain by Clyde Sandstone Formation, Ballagan Burn, Strathblane 13
4 a. Coarse agglomerate in funnel-shaped neck of vent cutting fine ash, Craigangowan Quarry, Milngavie 16
   b. South scarp of Campsie Fells showing trap features formed by lavas of the Clyde Plateau Volcanic Formation and the Lower Carboniferous vents, Dumfoyne and Dumgoyne 16
5 Stoop-and-room workings in the Baldernock Limestone (Lawmuir Formation), Linn of Baldernock 29
6 a. Shallow workings in the Giffnock Main Coal (Limestone Coal Formation) showing stoops and rooms, Thornliebank 36
   b. Stoop-and-room mining in the Bishopbriggs Sandstone (Upper Limestone Formation) Huntershill, Bishopbriggs, north Glasgow 36
7 Selected Carboniferous fossils 44
8 Fossil Grove, natural casts of lower rooted portions of *Lepidodendron* trees 45
9 Selected Devensian fossils 56–57
10 Wilderness Till Formation on sand and gravel of the Cadder Formation, Wilderness Sandpit, near Bishopbriggs 63
11 Core from the Killearn Borehole 75
   a. Seasonally laminated lacustrine sediments of the Blane Water Formation
   b. Glaciomarine sediments of the Paisley Formation
12 Landslip and scree on the flanks of Dumgoyne, a volcanic vent of Lower Carboniferous age, near Strathblane 77
13 The Whangie, landslipped mass of basalt showing deep, open back fissure, Auchineden Hill 77

PREFACE

An understanding of geology is essential to the sustainable development of the economy of the UK. Geology is particularly important in relation to the exploration for and development of mineral resources, the recognition of natural and manmade geological hazards, and their implications for land-use planning by national and local governments. In recognition of this the British Geological Survey is funded by central government to improve our understanding of the three-dimensional geology of the UK national domain through a programme of data collection, interpretation, publication and archiving. One aim of this programme is to ensure coverage of the UK land area with modern 1:50 000 scale geological maps and explanatory memoirs.

This memoir on the Glasgow district provides the first account of the geology of the western half and a modern account of the eastern half of an area which extends from Clydebank and Johnstone in the west to central Glasgow in the east and from Barrhead in the south to Killearn in the north. The northern part of the district, which is sparsely populated, has many features of considerable interest and importance in understanding the geological development of the Midland Valley. The central and south-eastern parts of the district which are among the most densely populated and industrialised in Scotland are rich in industrial archaeology, the coals, ironstones, fireclays and limestones of Carboniferous age having supported a thriving mining industry during the 19th and early part of the 20th centuries. Mining reserves became uneconomical in recent times, although there is potential for opencast mining of coal. Hard-rock aggregate is currently being utilised and there are viable resources of sand and gravel, and clay suitable for brick making.

A thorough knowledge of the stratigraphy described here and the location of old mine workings is essential for the planning necessary to facilitate the transition currently taking place in the area from heavy to light industry.

Peter J Cook, CBE, DSc, FGS, CGeol
Director

British Geological Survey
Kingsley Dunham Centre
Keyworth
Nottingham
NG12 5GG

ACKNOWLEDGEMENTS

This memoir was compiled by Mr I H S Hall and edited by Messrs M A E Browne and A D McAdam. Chapters in this memoir have been written by the following authors: **Introduction**, **Devonian** and **Carboniferous — general** by I H S Hall; **Lower Carboniferous** by I H S Hall (Inverclyde and Strathclyde groups) and I H Forsyth (Clackmannan Group); **Upper Carboniferous** by I H Forsyth; **Carboniferous palaeontology** by P J Brand: **Late- and post-Carboniferous intrusive igneous rocks** by I H Forsyth; **Structure** by I H S Hall (regional context) and I H Forsyth (folds and faults); **Quaternary** by M A E Browne; **Economic geology** by I H S Hall, I H Forsyth, M A E Browne and N S Robins; **Geophysical investigations** by M J Arthur.

Part of the section on the Clyde Plateau Volcanic Formation is based on notes and data from thin sections and analyses provided by Dr P M Craig. Dr D Stephenson contributed the section on the Beith–Barrhead Hills. The Carboniferous macrofauna were revised by Mr P J Brand and Dr R B Wilson, and the Carboniferous miospores were identified by Dr B Owens. The Quaternary macrofauna were identified by Mr D K Graham, the microfauna by Miss D M Gregory and Dr I P Wilkinson and the palynology by Miss C J Elliot and Mr R G M Newnham. Mr N S Robins contributed the hydrogeology for the Economic Geology chapter.

The cores and samples from boreholes, drilled by the Geological Survey in 1975–78 and 1983 to examine the Carboniferous succession and in 1979, 1981 and 1986 to investigate the Quaternary sediments, were examined by Messrs M A E Browne, I H Forsyth, I H S Hall, D N Halley, K I G Lawrie, A A McMillan and Dr S K Monro. The cores from boreholes drilled by civil engineering contractors to investigate foundation conditions and exploratory boreholes drilled by British Coal were examined by Messrs M A E Browne, J M Dean, I H Forsyth, I H S Hall, D N Halley and K I G Lawrie.

Photographs were mainly taken by Messrs T S Bain and F I MacTaggart.

NOTES

The word 'district' is used in this memoir to denote the area included in the 1:50 000 Geological Sheet 30E (Glasgow).

National Grid references are given in square brackets throughout this memoir; all lie within the 100 km square NS.

Numbers preceded by the letter S refer to the Sliced Rock Collection of the British Geological Survey.

Boreholes listed in Appendix 1 are held by the British Geological Survey, Murchison House, West Mains Road, Edinburgh, EH9 3LA.

HISTORY OF THE SURVEY OF THE GLASGOW SHEET

The Glasgow district is covered by Sheet 30E of the geological map of Scotland which was originally surveyed by J Geikie, E Hull and R L Jack and published at a scale of one inch to one mile in 1878. The area was resurveyed between 1904 and 1954 by B N Peach, L W Hinxman, C B Crampton, E B Bailey, E M Anderson, R G Carruthers, J E Richey, C H Dinham, J B Simpson, W G Kennedy, J G C Anderson, G S Johnstone and I H Forsyth and Solid and Drift maps were published in 1958 and 1961 respectively. The eastern half of the Glasgow district was published in 1911 as part of the Glasgow District Special Sheet. A second edition of the Special Sheet, which included additional information, was published in 1931 as separate Solid and Drift maps. Resurvey of the district was commenced in 1953, the solid being completed in 1985 and the drift in 1987. The survey was carried out by I H Forsyth, P M Craig, I H S Hall, M A E Browne, A A McMillan, D Stephenson, I B Paterson, A M Aitken and J M Dean under the supervision of J Knox, T R M Lawrie, J H Hull and J I Chisholm. Separate Solid and Drift editions of the Glasgow Sheet were published in 1993 at the scale of 1:50 000. The work was financed in part by the Department of the Environment between 1982 and 1985 during land-use-for-planning studies in Glasgow.

Neither of the previous surveys of the Glasgow district was accompanied by a sheet explanation, although the part which lies on the Glasgow Special Sheet was described in a special memoir of the Glasgow District, the 1st edition being published in 1911 and a 2nd edition in 1925. The memoirs describing the economic geology of the Central Coalfield of Scotland, Areas I, IV, V and VII also refer to the Glasgow district, although the most recent of these was published in 1937.

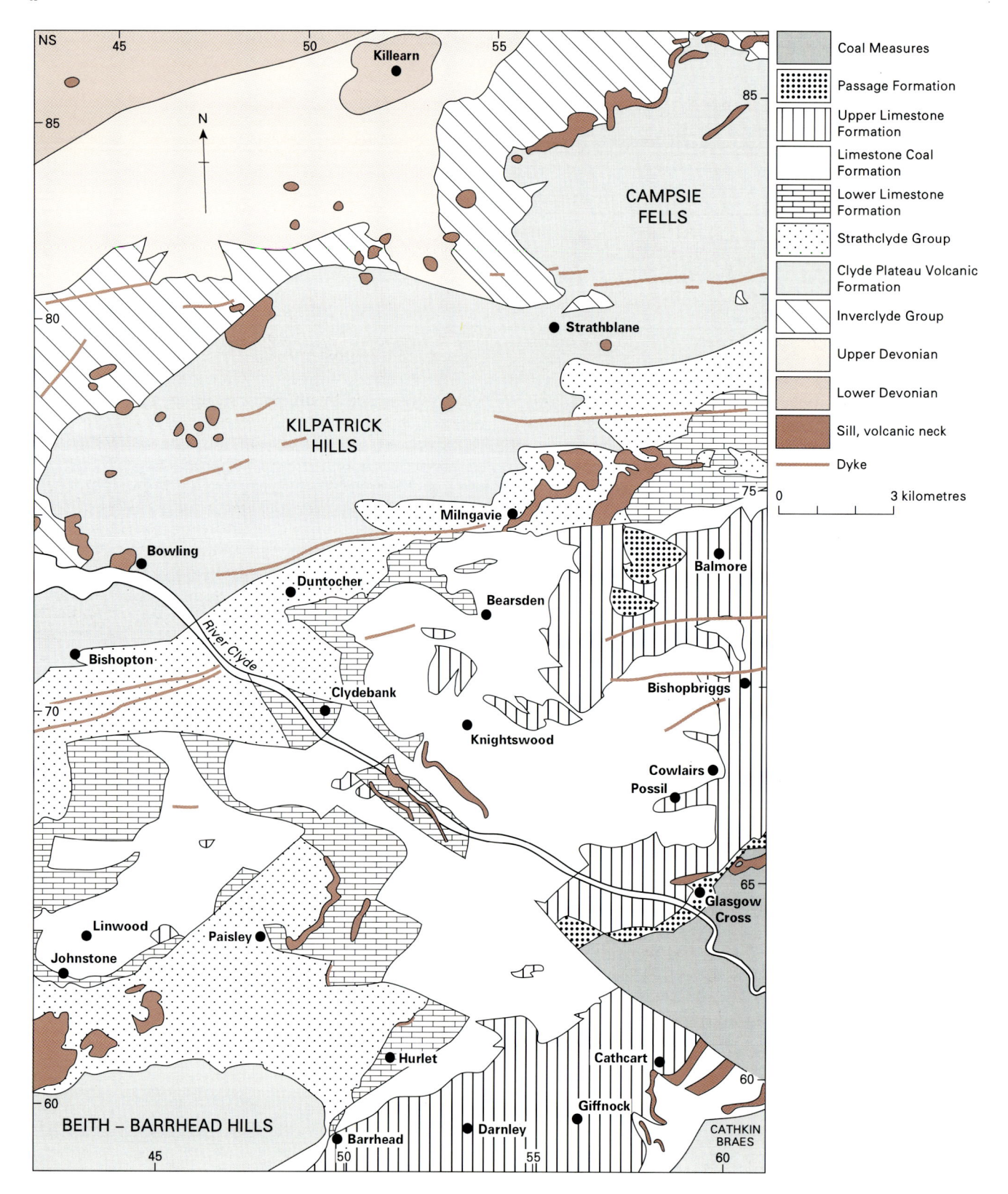

Figure 1 Geological map of the Glasgow district.

ONE

Introduction

The Glasgow district, which is described in this memoir, comprises the area covered by Sheet 30E of the 1:50 000 geological map of Scotland. The greater part of the district is occupied by a wide, gently undulating plain where Glasgow and most of the other urban development is situated. To the north, the ground rises to the dominant landscape features of the area, the Campsie Fells and Kilpatrick Hills. Together, they form a 6 km-wide tract of uplands extending across the district from the north-east corner to Bowling on the River Clyde in the west. The ground also rises along the southern margin of the district to the Beith–Barrhead Hills in the south-west and the Cathkin Braes in the south-east. Differential resistance to erosion of the different rock types is clearly reflected by the nature of the topography, almost all the features of prominent relief being due to the presence of extrusive or intrusive igneous rocks. The greater part of the uplands, both in the north and south of the district, are formed by erosion-resistant lavas and associated intrusions of the Lower Carboniferous Clyde Plateau Volcanic Formation (Figure 1). Most of these uplands form plateaux bounded by steep scarps, the exception being the Kilpatrick Hills where a pronounced dip to the south-east has allowed differential weathering of rocks within the lava pile to form a series of scarps with intervening dip slopes. Older strata ranging in age from Lower Devonian to Lower Carboniferous occur to the north-west of the Kilpatrick Hills and Campsie Fells. They consist of fluvial sandstones of the Teith, Stockiemuir Sandstone, Kinnesswood and Clyde Sandstone formations and the marginally marine mudstones with cementstones of the Ballagan Formation (Figure 2). Small areas of the latter also occur along the southern margin of the Campsie Fells. The lower ground south of the Campsie Fells and Kilpatrick Hills is occupied by a sequence of younger, less resistant, cyclic sedimentary rocks of Lower and Upper Carboniferous age. These strata consist of sandstones and mudstones with limestones, coals, ironstones and seatrocks which were laid down in fluvial and fluviodeltaic environments that were established after the submergence of the volcanic rocks. During the late Carboniferous and Permian these rocks were intruded by suites of quartz-dolerite and alkali dolerite sills and dykes which, in places, now give rise to prominent, low, flat-topped hills with abrupt escarpments and long, wall-like features. The undulating nature of the lower ground in the north-west part of the district is also partly due to the presence of numerous drumlins, sculpted largely in till by glaciers during the Quaternary.

The succession shows the effects of tectonic activity, being punctuated by unconformities and nonsequences. The most important of these occurred during the Middle Devonian when uplift and erosion took place and the regional palaeoslope was reversed from westward to eastward dipping. Another important break occurred in mid-Dinantian times when erosion of strata preceded the outpouring of the subaerial lavas of the Clyde Plateau Volcanic Formation. However, the presence of sandstones with erosive bases, notably near the top of the Limestone Coal Formation, and throughout the Upper Limestone and Passage formations, indicates intermittent tectonism during most of the Carboniferous.

During the Quaternary, the district was covered on several occasions by ice sheets which moulded the topography and laid down extensive deposits of till and glaciofluvial sand and gravel. The lower ground has also been subjected to several marine inundations during and since the last glaciation with the development of terraces and the deposition of estuarine sediments. Locally, basins, now commonly occupied by peat deposits, contained lakes e.g. the significant glacial lake that occupied Strath Blane about 11 000 to 10 000 years ago.

The frequent coal seams, ironstone beds and fireclays which occur in the Carboniferous sequence, and sand, gravel, clay and silt in the Quaternary deposits became the raw materials for the heavy industry which flourished in this area. Glasgow and several large towns including Clydebank, Paisley and Johnstone grew up round the centres of industry making the central part of the district one of the most densely populated areas in Scotland.

2 ONE INTRODUCTION

Figure 2 Generalised sequence of Devonian and Carboniferous rocks in the Glasgow district.

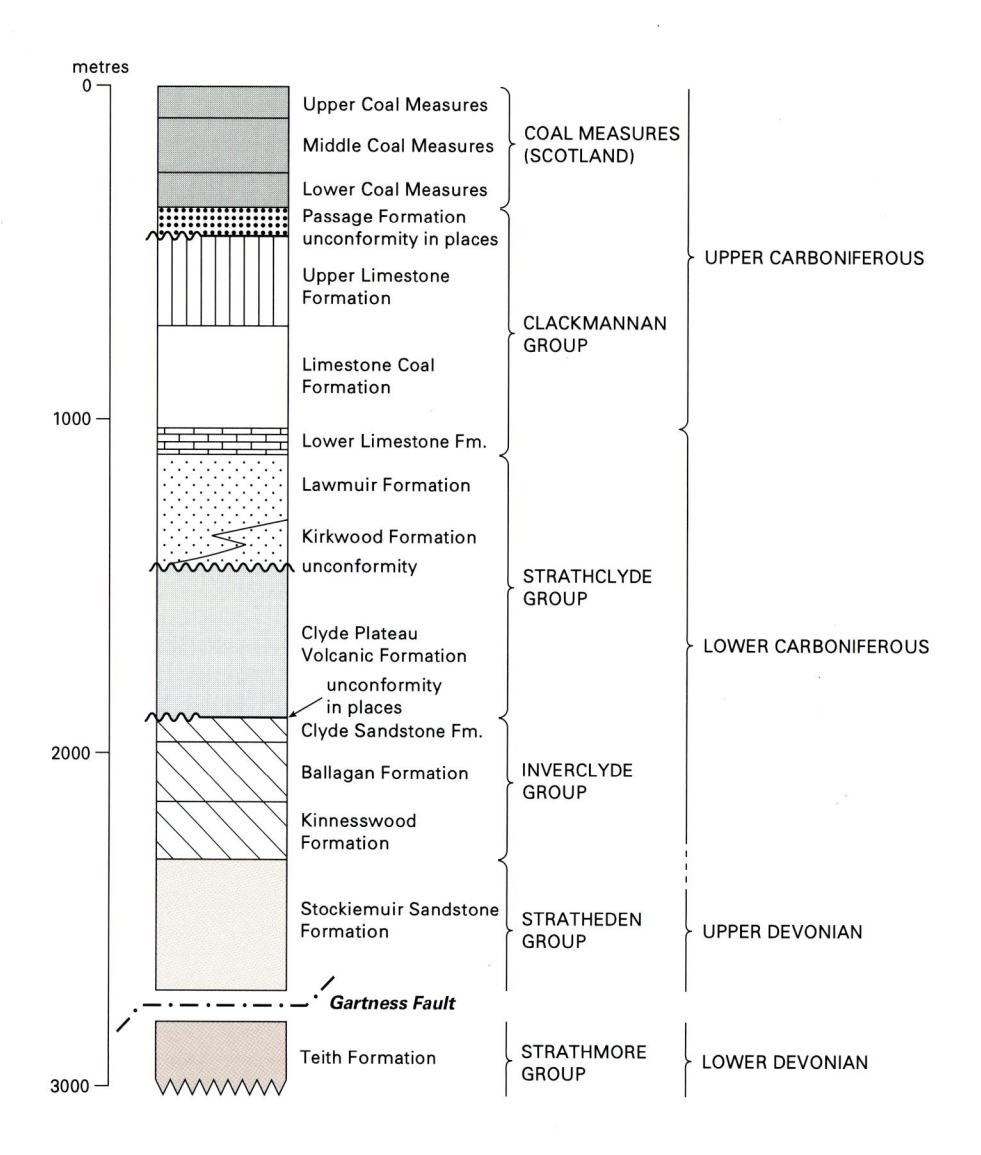

TWO
Devonian

Dominantly red and purple sandstones and siltstones that occupy about 60 km² of the lower ground to the north-west of the Kilpatrick Hills and Campsie Fells have been assigned to the Devonian. No stratigraphically significant fossils have been found in these rocks but they are correlated lithostratigraphically with parts of the Lower and Upper Devonian sequences of adjacent areas.

LOWER DEVONIAN

Lower Devonian rocks crop out along the north-west margin of the district where they are bounded to the south by Upper Devonian strata on the downthrown side of the ENE-trending Gartness Fault. The strata generally dip to the south-east or north-east and form part of the north-west limb and axial zone of the Strathmore Syncline, a major fold which extends across the whole of the Midland Valley. In the Glasgow district the south-east limb is concealed below younger rocks. Lower Devonian rocks also occur in an inlier around Killearn (Figure 1) where they are brought to the surface by gentle doming. The outcrops in the Glasgow district are part of a much larger area which has been assigned to the Teith Formation of the Strathmore Group (Francis et al., 1970). The very lowest part of the sequence, which is poorly exposed near Slatehouse [435 870], is probably equivalent to the upper part of the grey sandstone facies recognised in the Stirling district and the remainder can be correlated with the upper purple sandstone facies recognised in Stirling.

No identifiable fossils have been recorded in this district, but floras have been found in Teith Formation strata at several localities elsewhere in the Midland Valley which suggest a lower Emsian age (Paterson et al., 1990).

Strathmore Group

TEITH FORMATION

The oldest rocks occur in the north-west of the district near Slatehouse where there are a few outcrops of yellowish and brownish grey sandstones with scattered quartz pebbles. Some 50 m higher in the succession, about 40 m of red- and grey-purple sandstones are exposed east and north-east of Gartocharn [428 862]. These sandstones are arranged in small upward-fining units with pebbly bases and may be overlain by thin beds of red gritty mudstone. The pebbles are mostly of quartz, up to 5 cm in diameter, and red mudstone. Petrographically the sandstones are composed mainly of angular grains of quartz, feldspar and chert, with fragments of igneous rock, mudstone, and rarely mica-schist and mica. The cement is usually iron oxide but may be calcite. An analysis of a specimen from Gartocharn Quarry shows the silica content to be 76 per cent (Sabine et al., 1969). About 50 m higher in the sequence, strata consisting mainly of fine- to medium-grained purple-red 'ashy-looking' sandstones and mudstones are exposed in the Catter Burn west [468 858] and east [480 862] of Croftamie, where they are disturbed by the effects of the Gartness Fault. In this part of the sequence coarse-grained, pebbly sandstones are virtually absent. The 'ashy' appearance is caused by the presence of fine-grained volcanic detritus in the matrix. Strata at a slightly higher stratigraphical horizon are exposed in the valley of the Endrick Water at Gartness [502 868]. Here several upward-fining cycles occur with pebbly bases and thin red-purple mudstone preserved at the top. The pebbles are mostly composed of andesite and red mudstone. The Gartness Fault is exposed on the west bank [5026 8650].

Rocks of similar types occur around Killearn in an inlier surrounded by rocks of Upper Devonian age. Here the rocks are mostly purple-brown or reddish purple 'ashy-looking' sandstones with flaggy bedding. The purple-brown sandstones are coarse grained and commonly contain lenses of small well-rounded mainly andesitic but also rarely schist and quartz clasts. The reddish purple sandstones are finer grained. These are the youngest Lower Devonian strata occurring in the district and, though the unconformable junction with the overlying Upper Devonian is not exposed, proximity to it is marked by a more intense reddening of the Lower Devonian rocks.

Conditions of deposition

It is generally considered that deposition of the Lower Devonian sequence took place in an elongate, NE-trending basin, centred on the Midland Valley and bounded on the north-west by the Highlands (Bluck, 1978; Morton, 1979; Armstrong et al., 1985). Sediment derived from the Highlands was laid down along the northern basin margin in alluvial fans and subsequently partly dispersed south-westwards along the basin by axial drainage from the north-east. The fluvial sandstones and siltstones of the Teith Formation were deposited during one of the last phases known in the long and complex history of the Silurian-Devonian basin when the depositional environment was predominantly one of meandering streams on a broad floodplain.

UPPER DEVONIAN

Strata consisting mainly of red or red-brown sandstones, occurring in the north-west of the district, which were formerly referred to the Upper Old Red Sandstone, are now assigned to the Stratheden Group (Paterson and Hall, 1986). No fossils of stratigraphical significance have been found in these rocks in the Glasgow district but assemblages of fossil fish, which indicate the Famennian Stage of the Upper Devonian, have been found in similar strata elsewhere in the Midland Valley. The base of the Stratheden Group is not exposed in the Glasgow district. The top of the group is taken at the base of the succeeding cornstone-bearing sandstones of the Kinnesswood Formation, the lowest division of the Inverclyde Group. It is thought that the onset of the cornstone-forming facies took place near the Devonian–Carboniferous boundary and, since there is no faunal control, the top of the Stratheden Group is taken for convenience as the top of the Upper Devonian.

Stratheden Group

STOCKIEMUIR SANDSTONE FORMATION

All the Stratheden Group strata occurring within the district are assigned to the Stockiemuir Sandstone Formation (Paterson and Hall, 1986). The full thickness of the sequence cannot be determined because of faulting but is estimated to be more than 400 m. The lowest strata are seen around Killearn, south of Croftamie [477 858] and at Millfaid [460 854], and represent about 20 m of sequence. In all these sections the rocks are fine-grained red sandstones with small quartz and mica-schist pebbles and red mudstone intraclasts. Though millet seed grains are present in some beds (Hall and Chisholm, 1987), there are more features characteristic of water-laid deposits such as mica-covered surfaces and some parting lineations. The overlying 100 m of sequence is only poorly exposed. It is mainly represented by fine-grained well-bedded red sandstones, though a pebbly bed with quartz and mica-schist clasts up to 5 cm in diameter which is exposed near Lochend Cottage [430 841] probably lies within this interval. Above this, in the Carnock Burn [497 850 to 476 814], there is an almost continuous section through the remaining 300 m of sequence, up to the base of the overlying Kinnesswood Formation. The lowest 30 m of this part of the sequence, best seen in Finnich Glen [495 849] (Plate 1), are mainly of aeolian aspect (Hall and Chisholm, 1987) and comprise soft brick-red sandstones. These are partly cross bedded with downward-pinching non-laminated grain-flow wedges and partly flat-bedded sandstone in which 'pin-stripe' lamination of the type considered by Fryberger and Schenk (1988) to be characteristic of aeolian deposition, can be seen in bleached patches. Intercalations up to 5 m thick of water-laid fine-grained sandstones and siltstones with ripple-lamination and micaceous surfaces are present but these represent less than a quarter of the sequence. The same two facies occur in similar proportions throughout the overlying 270 m of section. Some strata of indeterminate facies are also present in the sequence. This is partly due to the 'pin-stripe' lamination characteristic of the aeolian facies being masked by the prevailing red colouration. In the eastern part of the area, extra-basinal pebbles are only present in the water-laid facies of the basal 20 m of the Stockiemuir Sandstone Formation. Further west near Gartlea [453 838], the Merkins [442 828] and on Blairquhomrie Muir [430 815], thin beds with quartz pebbles up to 5 cm in diameter occur at intervals in the water-laid sandstones throughout the sequence. In general, the water-laid facies here is coarser grained than around Croftamie.

Petrographically the sandstones are more mature than those of the Lower Devonian. They consist mainly of rounded to subangular quartz with some feldspar and chert and are cemented mainly by iron oxide. Two analyses of Upper Devonian sandstones from Dalreoch Quarry, Dumbarton, to the west of the Glasgow district,

Plate 1 Upper Devonian aeolian sandstone showing alternations of cross-laminated (dune facies) and parallel laminae (interdune facies), Finnich Glen, near Killearn (D 3765).

give 94 per cent silica compared with 76 per cent for a Lower Devonian sandstone (Sabine et al., 1969). The aeolian sandstones tend to be softer and less well cemented than the water-laid sandstones.

Palaeowind directions in the Croftamie area, based on the orientation of steep avalanche foresets, were measured by Hall and Chisholm (1987). The data show a bimodal pattern with a vector mean towards the south-east. In contrast, measurements taken in the Dumbarton area, near Stirling and in Fife can be interpreted to indicate dominant winds to be from easterly directions (Figure 3). The pattern is complicated by many reversals and cross winds.

Conditions of deposition

After a lengthy period of widespread earth movements accompanied by uplift and erosion, the regional Lower Devonian palaeoslope was reversed and Upper Devonian sedimentary rocks were laid down in a major ENE-trending depositional basin dominated by eastward-flowing rivers (Figure 3). During deposition of the Stratheden Group strata, the rainfall was probably moderate and somewhat seasonal. The presence of aeolian sandstones throughout the upper part of the Stockiemuir Sandstone Formation suggests that the rainfall decreased during this period. However, there would also be a predisposition to the development of aeolian sandstones since land vegetation was not as yet well developed. The water-laid sandstones in the western part of the district were probably laid down largely in channels on a extensive floodplain by meandering rivers. However, the finer-grained sandstones seen near Croftamie may represent the deposits of local short-lived flood basins. A return to exclusively fluvial sedimentation is seen in the overlying Kinnesswood Formation.

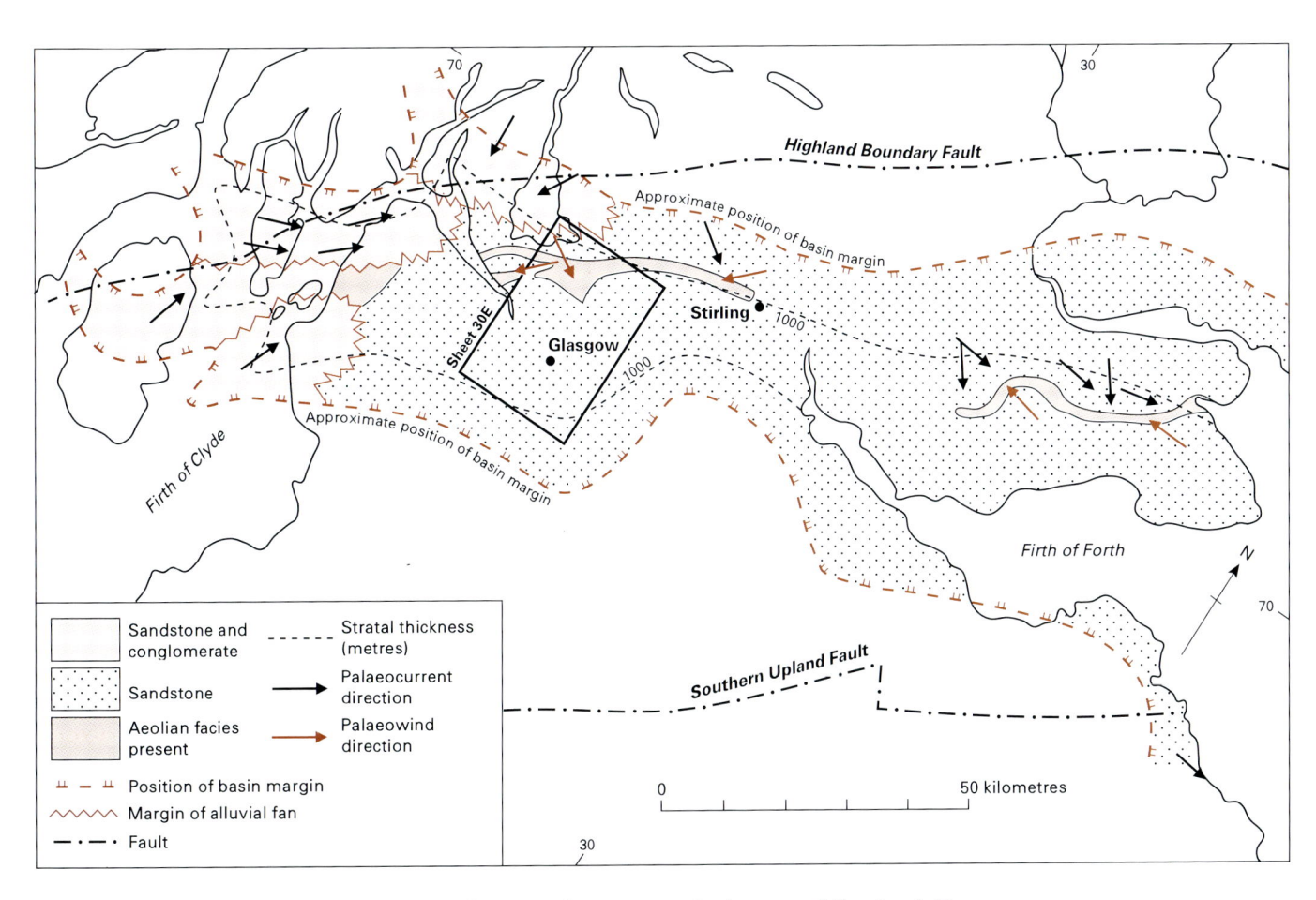

Figure 3 Reconstruction of Stratheden Group palaeogeography in central Scotland. Data partly from Bluck (1978; 1980), Chisholm and Dean (1974), Read and Johnson (1967) and Hall and Chisholm (1987).

THREE
Carboniferous — general

Apart from the area of Devonian in the north-west, strata of Carboniferous age crop out under the remainder of the Glasgow district. The oldest rocks thought to be of Carboniferous age are unfossiliferous, fluvial, mainly red and white sandstones with thin beds of concretionary limestone which rest on the Upper Devonian with a sharp junction in most of the district. They were deposited in an environment transitional between the arid or semi-arid conditions in which the generally red, fluvial and aeolian sandstones of the Upper Devonian were laid down and the humid environment in which the mainly grey, fluviodeltaic and marine Carboniferous strata were deposited. A return to semi-arid conditions occurred in the late Carboniferous when the region was uplifted and the Upper Coal Measures were oxidised and reddened. These climatic differences were due to the changing position of the Midland Valley relative to the equator, drifting from some distance south in the late Devonian to low latitudes in the early Silesian and well north by the end of the Carboniferous.

During the Dinantian, sedimentation was interrupted by a period of major subaerial volcanic activity during which over 400 m. of alkali basalts were erupted. These volcanic rocks have a surface area of over 150 km^2 in the district. They are thought to be present at depth under a considerable area occupied by younger Carboniferous sedimentary rocks.

CLASSIFICATION

A formal lithostratigraphical nomenclature has recently been applied to the late Devonian and early Carboniferous of the Midland Valley of Scotland (Paterson and Hall, 1986), whereby formation and group names have been erected and defined. Subsequently the group names have been applied to Fife (Browne, 1986) and new formations proposed. In the Lothians, the group names were also found to be valid (Chisholm et al., 1989) and further new formations were erected and defined. Since then lithostratigraphical criteria have been applied to the remainder of the Carboniferous (Forsyth et al., 1996).

The classification erected by Paterson and Hall (1986) included all the Lower Carboniferous up to the base of the Hurlet Limestone. Above this, the presence of widespread limestones, marine bands and coals allows correlations to be made across the Midland Valley on the basis of marker-band stratigraphy. All the units from the Hurlet Limestone upwards are defined by the bases or tops of limestones or the bases of marine bands. These are considered to be lithological markers defining the lithostratigraphical units. In accordance with the protocol established by the North American Commission on Stratigraphic Nomenclature (1983), these units are considered to be formations and have been assigned to recently established groups (Forsyth et al., 1996). Since the definitions of lower and upper boundaries have not been changed, previous descriptions are still valid. In view of the general familiarity with longstanding names, these have have been modified as little as possible in the new classification. Thus, the Lower Limestone Formation replaces the term Lower Limestone Group.

Lithostratigraphy of the Lower Carboniferous (Table 1)

Classification of all except the highest part of the Lower Carboniferous has been discussed, type sections erected and formal names defined by Paterson and Hall (1986) and Chisholm et al. (1989). This part of the succession includes all the strata up to base of the Hurlet Limestone, traditionally taken as the top of the former Calciferous Sandstone Measures. The overlying dominantly argillaceous cyclic sequences with sandstones, generally thin limestones and a few coals, which occur between the base of the Hurlet Limestone and the top of the Top Hosie Limestone, has been renamed the Lower Limestone Formation and assigned to the Clackmannan Group (Forsyth et al., 1996).

Lithostratigraphy of the Upper Carboniferous (Table 1)

The Upper Carboniferous starts with the Limestone Coal Formation. The lower part of this formation comprises a sequence of dominantly argillaceous cyclothems with sandstones, mudstones, seatrocks, coals and ironstones. The upper part of the formation consists of dominantly arenaceous cyclothems. Two marine bands and several *Lingula* bands are present but there are few marine limestones. Stratigraphically above, is a succession of dominantly arenaceous cyclothems with mudstones, marine limestones and thin coals forming the Upper Limestone Formation. The succeeding strata, which are mainly fluvial in origin with thin marine bands, comprise the Passage Formation. The overlying fluviodeltaic coal-bearing sequence is divided into Lower, Middle and Upper Coal Measures at the Vanderbeckei (Queenslie) and Aegiranum (Skipsey's) marine bands respectively. The Lower Limestone, Limestone Coal, Upper Limestone and Passage formations have been assigned to the Clackmannan Group. The Lower, Middle and Upper Coal Measures are considered to be informal units with formation status and are placed in the Coal Measures, which has group status (Forsyth et al., 1996).

Table 1 Classification of the Carboniferous in western Scotland.

Subsystem	Series	Stage	Lithostratigraphical units	
			Formation	Group
Upper Carboniferous (Silesian)	Westphalian	Bolsovian (Westphalian C)	Upper Coal Measures *Aegiranum Marine Band*	Coal Measures
		Duckmantian (Westphalian B)	Middle Coal Measures *Vanderbeckei Marine Band*	
		Langsettian (Westphalian A)	Lower Coal Measures *Lowstone Marine Band*	
	Namurian	Yeadonian Marsdenian Kinderscoutian Alportian Chokerian	Passage Formation	Clackmannan Group
		Arnsbergian	*Castlecary Limestone* Upper Limestone Formation	
		Pendleian	*Index Limestone* Limestone Coal Formation	
Lower Carboniferous (Dinantian)	Viséan	Brigantian	*Top Hosie Limestone* Lower Limestone Formation *Hurlet Limestone*	
			Lawmuir Formation	Strathclyde Group
		Asbian Holkerian Arundian Chadian	Kirkwood Formation	
			Clyde Plateau Volcanic Formation	
	Tournaisian	Courceyan	Clyde Sandstone Formation	Inverclyde Group
			Ballagan Formation	
			Kinnesswood Formation	

Chronostratigraphy and biostratigraphy

The Carboniferous is divided into two subsystems, the Dinantian (Lower Carboniferous) and Silesian (Upper Carboniferous). The base of the Dinantian is internationally defined by the presence of particular marine fossils (Paproth et al., 1991). These fossils have not been found in Scotland and consequently the base cannot be precisely defined. Miospores are the only fossils of any biostratigraphical significance which occur in the lowest Carboniferous in the district. The Ballagan Formation is placed in the CM miospore zone (Neves et al., 1972), the oldest that can be recognised in Scotland. At least two lower zones are recognised elsewhere and these are thought to be present in the underlying unfossiliferous Kinnesswood Formation. Hence, at least part of this formation is considered to be of early Carboniferous age.

The base of the Silesian is taken about 1 m below the Top Hosie Limestone at the lowest recorded occurrence of the goniatite *Cravenoceras*. The Silesian is divided into the Namurian and Westphalian series which in turn are subdivided into stages, though only some of the diagnostic goniatites are present in Scotland. The base of the Westphalian is defined by the presence of *Gastrioceras subcrenatum*. This goniatite has not been found in Scotland and the base of the Lower Coal Measures has been taken at a suitable arbitrary position. Within the Glasgow district this is the Lowstone Marine Band (Table 1). Miospore evidence (Neves et al., 1965) suggests, however, that one of the marine bands in No. 6 Marine Band group (not known in the district) in the Passage Formation is the equivalent of the *G. subcrenatum* Marine Band and consequently, that the base of the Westphalian may lie within the upper part of this formation.

FOUR
Lower Carboniferous (Dinantian)

The Lower Carboniferous strata are assigned to three groups. These are the Inverclyde Group, the Strathclyde Group and the lowest part (the Lower Limestone Formation) of the mainly Upper Carboniferous Clackmannan Group (Table 1).

INVERCLYDE GROUP

The Inverclyde Group consists of the Kinnesswood, Ballagan and Clyde Sandstone formations (Figures 2 and 4, Table 1). These formations are exposed along the northern margin of the Campsie Fells and the Kilpatrick Hills, along the flanks of Strath Blane and around Bowling. They are not exposed south of the River Clyde within the district.

Kinnesswood Formation

The strata of this formation comprise commonly variegated, purple-red, yellow, white and grey-purple sandstones with nodules and thin beds of concretionary carbonate (cornstone) and a few thin beds of red mudstone. No fossils have been found in these strata within this district but to the west, at Fairy Knowe Quarry [369 789] near Dumbarton, disarticulated scales of *Bothriolepis* and *Holoptychus* of Famennian age have been found in the basal conglomerate (Aspen, 1974). However, general considerations suggest that the Kinnesswood Formation is mostly of early Carboniferous age (Paterson and Hall, 1986). These rocks were laid down on a broad alluvial plain by a generally eastward-flowing river system which occupied a large part of the Midland Valley at this time. The cross-bedded sandstones were deposited in the river channels and the cornstones formed as carbonate nodules in soil profiles that developed on the associated floodplains under the influence of a fluctuating water table in a semi-arid climate. Since argillaceous overbank deposits are rarely present in this district, the cornstones occur in sandstones at the top of channel-fill sequences and formed after the channels had been abandoned. The cornstones range from immature, in which the sandstones have a partly carbonate matrix with ill-defined isolated concretions, to mature, in which well-defined nodules (glaebules) are elongated in a vertical sense and are overlain by laminar and pisolitic structures (Plate 2). The laminar structures, which develop subparallel to the bedding, may be brecciated and the carbonate replaced by chert. Where silicification occurs, the fabric of the cornstone is preserved and the mineralogical change appears to be an integral part of the pedogenic process. Analyses of cornstones from just to the west of the district (Table 2, Figure 5) suggest that the carbonate is largely dolomite, probably secondary, though limestone is also known to occur. Since there is very little sediment deposition during the formation of cornstone profiles, it is unlikely that there was sufficient meteoric water to supply enough magnesium for the formation of primary dolomite. There is no evidence of a potential source in the underlying strata; the only other source is seawater. It also seems unlikely that the argillaceous nature of the

Plate 2 Partly silicified mature cornstone in the Kinnesswood Formation, Roughting Burn, near Dumbarton (D 3378).

Figure 4 Correlation of the Inverclyde Group in the Glasgow district.

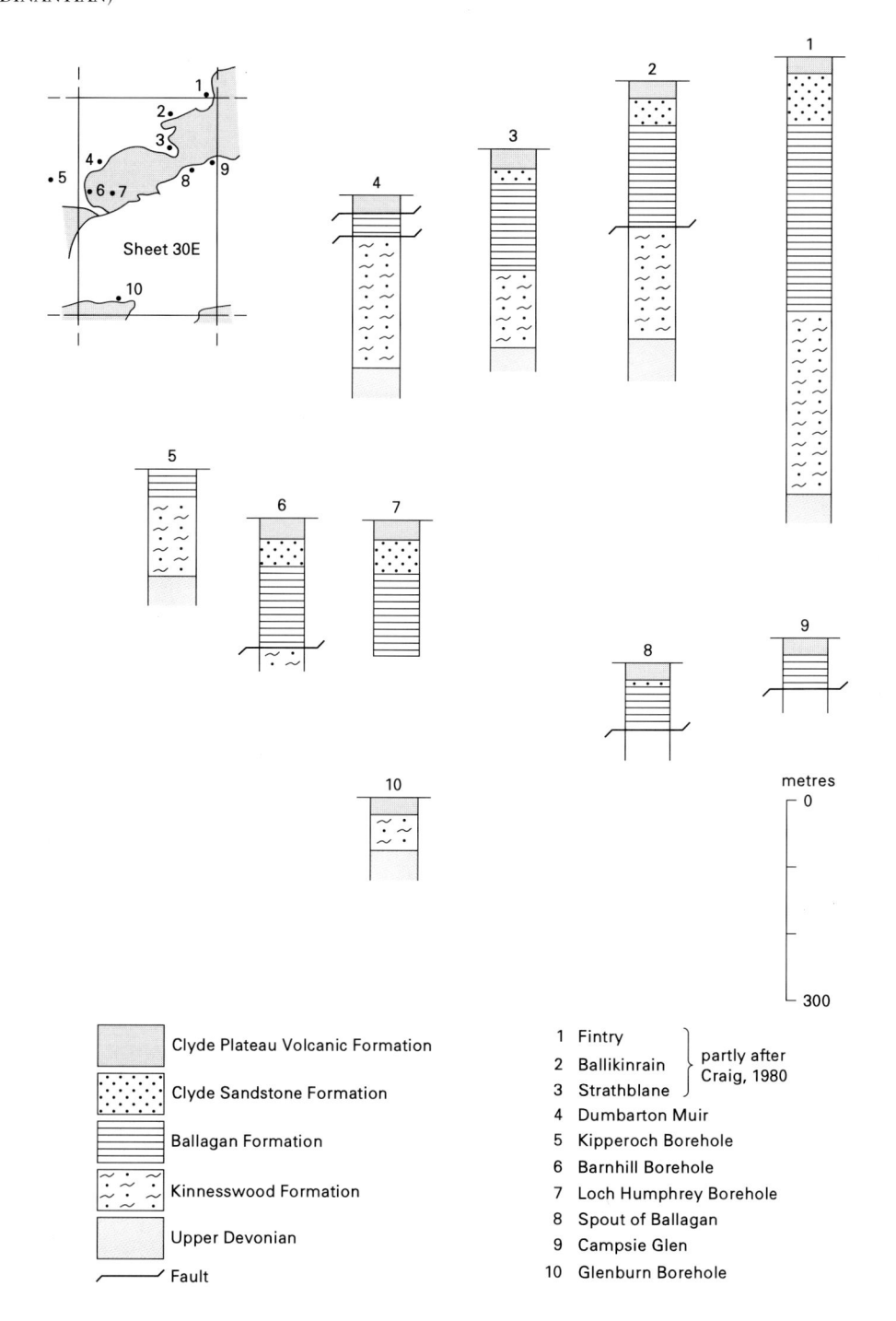

Clyde Plateau Volcanic Formation
Clyde Sandstone Formation
Ballagan Formation
Kinnesswood Formation
Upper Devonian
Fault

1 Fintry ⎱ partly after
2 Ballikinrain ⎰ Craig, 1980
3 Strathblane
4 Dumbarton Muir
5 Kipperoch Borehole
6 Barnhill Borehole
7 Loch Humphrey Borehole
8 Spout of Ballagan
9 Campsie Glen
10 Glenburn Borehole

overlying Ballagan Formation would allow sufficient groundwater to percolate downwards, particularly after lithification, to effect the secondary dolomitisation of the cornstones. However, mature cornstones develop over a lengthy period (up to 1 million years, Machette, 1985) during which they are exposed at the surface under a regime of low enough rainfall to allow mostly calcium carbonate to be deposited. A high rate of evaporation may have caused lateral movement of magnesium-rich groundwater, derived either directly from saline water in marginal marine areas where early Ballagan Formation strata were being deposited or indirectly from unlithified sediments during their early diagenesis, to effect dolomitisation of the cornstone profiles. A good example of a mature cornstone profile 2 m thick and traceable for 0.5 km occurs at Overton Burn [437 777].

Outcrops of the formation only occur north of the River Clyde and are largely confined to two areas. One is an area of high moorland north of the Kilpatrick Hills and the other lies on the eastern side of Strath Blane extending north from Blanefield to the northern margin of the district. Faulting makes accurate assessment of thickness of the formation across the district difficult but estimated values are shown in Figure 4. In general they ap-

Table 2 Determinations of various metals in the acid-soluble (carbonate) fraction of carbonate-rich rocks of the Stratheden and Inverclyde groups. Cation analyses are given as a percentage.

* From the Greenock district, after Paterson et al., 1990.

LISC = Lithostratigraphic code: BGN = Ballagan Formation, CYD = Clyde Sandstone Formation, KNW = Kinnesswood Formation, SCK = Stockiemuir Sandstone Formation.

Note: in a pure dolomite, the Ca/Mg ratio would be 1.67.

Sample No.	Depth (m)	LISC	Ca	Mg	Fe	Mn	Sr	Ba	Ca/Mg
Murroch Burn [4175 7870]									
FX 196	0	BGN	20.1	10.6	0.50	0.28	0.03	0.012	1.9
Overton Burn [4328 7758]									
FX 197	0	BGN	21.2	9.4	0.52	0.37	0.30	0.078	2.3
Barnhill Borehole [4269 7571]									
FX 198	153.24	CYD	29.3	1.2	0.55	0.061	0.06	0.003	24.4
FX 199	156.43	CYD	31.7	0.9	0.7	0.103	0.026	0.002	35.2
FX 200	170.07	BGN	16.6	10.3	1.12	0.133	0.24	0.053	1.6
FX 201	170.30	BGN	18.2	10.9	0.96	0.13	0.082	0.018	1.6
FX 202	181.62	BGN	18.2	11.1	1.05	0.17	0.050	0.013	1.6
FX 203	211.52	BGN	17.5	9.9	0.95	0.19	0.28	0.067	1.7
FX 204	228.88	BGN	18.5	9.7	0.83	0.19	0.37	0.04	1.9
FX 205	241.76	BGN	19.9	10.3	0.65	0.18	0.68	0.004	1.9
FX 206	249.22	BGN	17.1	9.6	0.97	0.14	0.15	0.022	1.7
FX 207	261.63	BGN	19.7	9.8	0.75	0.20	0.44	0.113	2.0
FX 208	270.17	BGN	17.3	9.8	0.71	0.24	0.047	0.017	1.7
FX 209	270.30	BGN	20.6	11.1	0.36	0.2	0.089	0.1	1.8
FX 210	282.82	BGN	16.9	9.5	0.78	0.19	0.015	0.002	1.7
Kipperoch Borehole [3727 7742]*									
FX 211	10.53	BGN	19.1	10.6	0.5	0.32	0.032	0.009	1.8
FX 212	13.47	BGN	17.5	10.4	0.61	0.22	0.115	0.014	1.6
FX 213	27.28	BGN	13.6	6.4	0.60	0.18	0.017	0.001	2.1
FX 214	27.55	BGN	14.0	8.1	0.74	0.18	0.015	0.18	1.7
FX 215	49.04	BGN	6.3	1.1	1.08	0.017	<0.005	0.003	5.7
FX 216	50.90	BGN	15.8	9.4	0.51	0.13	0.015	0.002	1.6
FX 217	51.33	KNW	12.5	7.7	0.55	0.096	0.045	0.003	1.6
FX 218	74.53	KNW	13.7	10.3	0.31	0.131	0.03	0.005	1.3
FX 219	77.20	KNW	17.2	8.7	0.50	0.11	0.01	0.002	1.9
FX 220	95.87	KNW	18.4	10.4	0.09	0.15	0.011	0.001	1.7
FX 221	121.62	KNW	15.9	8.4	0.076	0.086	0.007	<0.001	1.8
FX 222	121.91	KNW	12.8	6.9	0.055	0.067	0.006	0.002	1.8
FX 223	122.01	KNW	13.4	6.8	0.054	0.064	0.006	0.003	1.9
FX 224	122.20	KNW	11.4	6.5	0.057	0.06	0.006	0.003	1.7
FX 225	192.80	KNW	13.9	7.3	0.114	0.136	0.006	0.003	1.9
FX 226	217.96	KNW	16.3	4.3	0.061	0.17	0.007	0.005	3.7
FX 227	223.58	SCK	3.5	0.1	0.046	0.034	0.003	0.002	35.0
FX 228	231.36	SCK	3.8	0.1	0.054	0.042	0.003	0.005	38.0
FX 229	234.10	SCK	7.6	0.3	0.049	0.072	0.005	0.003	25.3
FX 230	279.26	SCK	2.6	0.1	0.073	0.043	0.003	0.025	26.0
Everton Borehole [2145 7103]*									
ZH 1267	22.35	BGN	17.7	8.0	0.8	0.2	0.031	0.006	2.2
ZH 1268	36.70	BGN	23.1	9.9	0.24	0.22	0.052	0.001	2.3
ZH 1269	39.35	BGN	17.4	8.1	0.65	0.15	0.039	0.16	2.1
ZH 1270	43.60	BGN	14.9	8.2	1.07	0.12	0.026	0.005	1.8
ZH 1271	108.70	KNW	31.8	0.44	0.084	0.026	0.026	0.002	72.2
ZH 1272	113.85	KNW	26.5	1.24	0.115	0.062	0.025	0.001	21.3
ZH 1273	120.80	KNW	29.6	0.38	0.076	0.039	0.035	0.002	77.8
Knocknairshill Borehole [3056 7438]*									
ZH 1274	138.90	CYD	29.9	0.33	0.13	0.1	0.041	0.018	90.6
ZH 1275	154.90	CYD	21.3	0.41	0.31	0.23	0.025	0.005	51.9
ZH 1276	173.45	CYD	17.4	0.72	0.39	0.67	0.024	0.002	24.1
ZH 1277	181.90	CYD	23.8	0.66	0.14	0.06	0.033	0.022	36.0
ZH 1278	188.13	CYD	15.3	6.8	0.24	0.032	0.43	0.006	2.2
ZH 1279	224.70	BGN	18.1	8.9	0.89	0.17	0.026	0.012	2.0
ZH 1280	241.40	BGN	32.8	0.91	0.36	0.115	0.033	0.003	36.0
ZH 1281	257.10	BGN	30.0	1.24	0.51	0.115	0.096	0.009	24.1
ZH 1282	272.45	BGN	22.2	9.5	0.27	0.2	0.5	0.016	2.3
ZH 1283	304.90	KNW	25.4	0.82	0.15	0.15	0.032	0.008	30.9
ZH 1284	330.35	KNW	20.0	5.3	0.29	0.19	0.023	0.002	3.7
ZH 1285	347.25	KNW	4.7	0.77	0.012	0.055	<0.005	0.035	6.1
ZH 1286	379.40	KNW	11.0	5.8	0.078	0.26	<0.005	0.013	1.9

Figure 5 Molar percent CaCO$_3$ in carbonates.

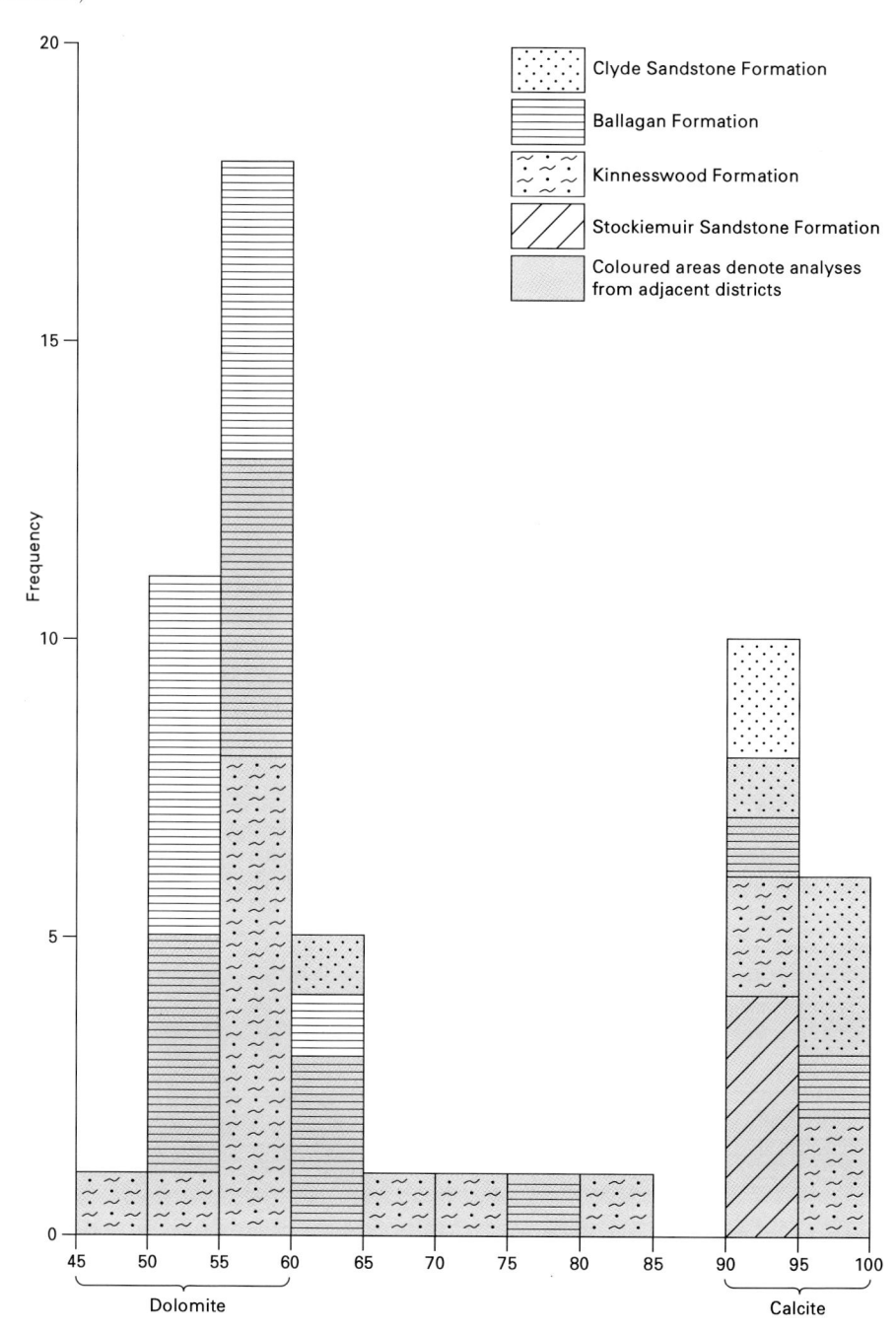

Legend:
- Clyde Sandstone Formation
- Ballagan Formation
- Kinnesswood Formation
- Stockiemuir Sandstone Formation
- Coloured areas denote analyses from adjacent districts

pear to be greater than the 170 m which was encountered in the Kipperoch Borehole [3727 7742] to the west of the district, where a complete sequence was drilled.

On Dumbarton Muir, north of the Kilpartick Hills, good exposures of Kinnesswood Formation occur, mostly in stream sections. In this area the formation reaches more than 250 m but is variable in thickness and character. The section in and adjacent to Finland Burn [450 818–432 800] is comprised of poorly bedded, fine- to medium-grained grèy- and red-purple sandstones interbedded with paler, coarser-grained cross-bedded sandstones. Many of the finer-grained sandstones contain carbonate concretions that may pass up into cornstones, though the latter are less common in the upper part of the sequence. The coarser-grained sandstones, which are interpreted as channel-fill deposits, have sharp or erosional bases and are frequently cross bedded. In the lower part of the sequence these sandstones may have cornstone clasts near their bases, whereas in the upper part they frequently have quartz pebbles up 10 cm in diameter. The finer-grained sandstones may be overbank deposits or channel deposits which have lost their original structures when soil profiles developed in the the fill of abandoned channels. The cornstones are 0.3–1 m thick and some show partial silicification. Several of these were quarried and burnt locally. Millstones were quarried from one of the coarser sandstones on Dumbarton Muir [444 804]. In contrast, less than 2 km to the east, in the Gallangad [455 814–450 783] and Knockupple [452 802–458 786] burns, cornstones are rare in the lower part of the sequence but well developed in the upper part. Quartz pebbles also occur near the bases of the coarser-grained sandstones throughout the sequence. Farther east on Stockiemuir [48 81] and on the west side of Strath Blane, cornstones are not seen and fine-grained sandstones, where present, show few concretionary textures.

On the eastern side of Strath Blane, good sections of the Kinnesswood Formation are exposed in several streams draining off the Campsie Fells. In this area the formation is about 170 m thick near Blanefield and thickens northward to 400 m near Fintry at the western end of the Gargunnock Hills (Francis et al., 1970). At the northern end of this area on Ballikinrain Muir [56 86], the formation is in two generally upward-fining sequences. The lower cycle, which is about 280 m thick, comprises coarse-grained, pink and purple-red, pebbly sandstones interbedded with grey- and red-purple fine-grained sandstones and cornstones. The coarser-grained sandstones become less frequent and the cornstones better developed higher in the sequence. Here, the pebbles are of quartz, quartzite and schist. The upper cycle, which is 50 m thick, consists of white, coarse-grained siliceous sandstone with lenses of small quartz pebbles, overlain by pale grey, fine-grained sandstones interbedded with subordinate greenish siltstones and cornstones. Several sandstones were formerly quarried for millstones in this area. Both cycles change in character when traced southward towards Blanefield. The coarse pebbly facies at the base of the lower cycle is replaced by about 60 m of pale purplish or greenish grey, medium-grained, massive, cross-bedded or flaggy sandstone. This is overlain by alternations of red-purple coarse-grained sandstones with quartz pebbles and finer-grained sandstones with cornstones, the latter becoming more frequent near the top of the cycle. The upper cycle commences with 10 m of white coarse gritty, siliceous sandstone with frequent quartz pebbles up to 15 cm in diameter and is followed by 20 m of fine- to medium-grained white sandstone, commonly calcareous, with mature cornstones and a few interbeds of greenish grey mudstone.

South of the River Clyde, a complete sequence was drilled in the BGS Glenburn Borehole [4783 6066] where the thickness was found to be 73 m. Here, mature cornstones are not well developed, though some small carbonate concretions are present at intervals throughout the lower 40 m. Mudstone clasts and small quartz pebbles are present throughout the sequence, usually being more abundant at or near the base of sandstone units. Many of the medium- and coarse-grained sandstones are cross-laminated, fine upwards and have erosive bases, features typical of channel-fill deposits.

Ballagan Formation

The succeeding Ballagan Formation, named after the type section in Ballagan Glen [510 800] (Plate 3), is well exposed at the western end of the Kilpatrick Hills and along the western and northern flanks of the Campsie Fells. Between these two areas the formation is largely cut out by faulting. The sequence in the two areas is very similar. Typically, the strata consist of grey, poorly laminated silty mudstones with frequent thin nodular beds of dolomitic limestone (cementstone) and a few thin, fine-grained sandstones occurring at intervals throughout the sequence. At the base of the formation, more frequent sandstone beds are intercalated among the mudstones in a zone up to 20 m thick, which is transitional with the underlying Kinnesswood Formation. Reddening, which is normally associated with the sandstones, can also affect the mudstones and cementstones locally.

South of the River Clyde, the Glenburn Borehole [4783 6066] proved the Clyde Plateau Volcanic Formation lying directly on the Kinnesswood Formation (Figure 4), strata of the Ballagan Formation and the

Plate 3 Alternating beds of mudstone and cementstone of the Ballagan Formation overlain by sandstone of the Clyde Sandstone Formation, Ballagan Burn, Strathblane (D 1883).

overlying Clyde Sandstone Formation having been removed by erosion during the mid-Dinantian (Forsyth et al., 1996).

The cementstones occur as thin beds of nodular carbonate or as layers of discrete ellipsoidal nodules elongated parallel to the bedding. Though rarely exceeding 0.2 m in thickness, individual beds may persist laterally for distances of over 2 km. Most have no apparent internal structure but some are brecciated, and a few are laminated. Septarian cracks commonly occur and are mostly infilled with crystalline dolomite which rarely is accompanied by copper and lead minerals. Cavities are common and may be empty or lined with gypsum. In one case the gypsum was encrusted with copper minerals. All the cementstones analysed from this district were found to be dolomitic (Table 2; Figure 5), as was the case in adjacent districts (Paterson et al., 1990; Francis et al., 1970). The occurrence of evaporitic minerals (see below), and the lack of other potential sources, points to the magnesium in the dolomite being derived from seawater which has been concentrated by evaporation in restricted saline lakes. It has been suggested that the laminated cementstones were deposited as carbonate muds (Belt et al., 1967) and that the nodular beds are diagenetic in origin, probably formed early and at shallow depth (Andrews et al., 1991).

The mudstones are normally grey, poorly laminated and may have desiccation cracks. They are patchily calcareous in places and, where seen in borehole core, have frequent gypsum nodules and are cut by numerous veins of fibrous gypsum up to 8 cm thick. It is noticeable that much more gypsum is present in strata drilled in boreholes (e.g. Barnhill and Loch Humphrey) than in nearby exposures. This is thought to be due to gypsum having been dissolved out by recent surface weathering. The abundance of gypsum veins may be due to the conversion of anhydrite to gypsum since, as suggested by Shearman et al. (1972), this change involves an incease in volume of 63 per cent in the solid phase and gypsum veins occurring in association with gypsum nodules represent, at least in part, the additional volume created by this conversion. The veins were considered to be emplaced by hydraulic fracture. Also seen in the cores are brecciated zones which may have been caused by collapse after the solution of thin salt deposits (? halite and anhydrite). Halite pseudomorphs do occur at intervals in the mudstones and are also present in two of the fine-grained sandstones. Scott (1986) recorded textural and mineralogical features from the Ballagan Formation of Berwickshire which he interpreted as evidence for a former sulphate evaporite facies. He also pointed out that the deposition of gypsum would reduce the amount of dissolved calcium thus increasing the relative amount of magnesium in the brine and promoting the requisite Mg/Ca ratio for the formation of dolomite. A little pyrite and plant debris are also present in the mudstones.

Thin fine-grained sandstones which may show ripple cross-lamination or cross-bedding, occur at intervals. A few are very micaceous and poorly cemented but the majority are more massive and have a carbonate cement. The cross-laminated sandstones are small channel deposits, the remainder may be sheet-flood deposits.

The only fossils which have been found in the Ballagan Formation in the district are scattered occurrences of '*Estheria*' sp. and ostracods which were recorded in BGS Barnhill [4269 7571] and Loch Humphrey [4592 7555] boreholes, and *Spirorbis* sp. and ostracods in surface exposures.

The dolomitic cementstones are thought to have been deposited in coastal sabkhas, restricted saline lakes or lagoons frequently replenished by seawater. The frequent repetition of cementstones interbedded with mudstones, siltstones and sandstones, deposited as terrigenous sediments on a mature coastal alluvial plain dominated by mudflats, forms the characteristic element of a sequence of strata unique in the Scottish Carboniferous.

Clyde Sandstone Formation

A return to a fluviatile environment is seen in the Clyde Sandstone Formation, the youngest division of the Inverclyde Group. In the Glasgow district, outcrops of this formation occur in geographically isolated areas north of the River Clyde. Good exposures occur at the western end of the Kilpatrick Hills and at intervals flanking the Campsie Fells with only a few fault-bounded outcrops between. South of the River Clyde, in the Gleniffer Braes area, the formation has been removed by mid-Dinantian uplift and erosion (Forsyth et al., 1996).

The sandstones are coarser grained than those of the Kinnesswood Formation and though commonly calcareous and concretionary, only locally develop into cornstones. Analyses (Table 2; Francis et al., 1970) and field examination suggest that all the carbonate is calcite rather than the dolomite which characterised the underlying formations. In the more northerly areas the sandstones are conglomeratic with quartz and Highland pebbles in places. Elsewhere, the pebble content is dominated by intraformational clasts of mudstone or limestone. Farther south there are no clasts present. Thus the strata of this formation represent the deposits of a wide variety of fluviatile environments, ranging from braided stream to floodplain with well-developed overbank.

West of the Kilpatrick Hills the formation is represented by the Overtoun Sandstone Member. The complete sequence was penetrated by the Barnhill Borehole where it was 57 m thick. The lower part is dominated by grey, fine-grained sandstones which are thin bedded or partly concretionary. There are also some grey mudstones which contain bands or concretions of limestone, and erosive-based coarser-grained sandstones with carbonate clasts. The upper part consists largely of sandstones with erosional bases and containing carbonate clasts; concretionary sandstones and mudstones are uncommon. Exposures occur in the Overtoun Burn [423 760].

Between this area and Strath Blane, faulted outcrops occur on Auchineden Hill [485 802, 497 809] and Dumgoyach Brae [525 806]. On Auchineden Hill, about 6 m of white gritty sandstone are exposed, containing abundant quartz pebbles up to 6 cm in diameter. On Dumgoyach Brae there is a 12 m sequence of brownish trough cross-bedded sandstone with bands of quartz pebbles; some of the sandstones are concretionary.

Exposures of the formation also occur along the northern and western flanks of the Campsie Fells. However, on the south side, in Campsie Glen [610 800] and Fin Glen [600 799], lavas of the overlying Clyde Plateau Volcanic Formation lie directly on the Ballagan Formation, strata of the Clyde Sandstone Formation having been removed during the mid-Dinantian (Forsyth et al., 1996). In Ballagan Glen [572 801], to the west of these localities, the formation is about 12 m thick but the section is now largely inaccessible. It consists of white siliceous sandstones interbedded with siltstones and mudstones. A seatearth with numerous coal streaks was formerly seen (Macgregor et al., 1925) near the top of the section. Clasts appear to be absent in this area. Along the east side of Strath Blane, the formation is poorly exposed but farther north there is a good section in Little Corrie [576 844] where the sequence is about 45 m thick. The lower part consists of coarse-grained pebbly sandstone passing up into commonly ripple-laminated, micaceous, finer-grained sandstones, grey siltstones and mudstones with desiccation cracks. Some of the sandstones are concretionary with a calcareous matrix. The bulk of the clasts are quartz with some quartzite. This is in marked contrast to the sandstones to the north-east in the Stirling district where the clasts are entirely of limestone (Read and Johnson, 1967).

The distribution and types of clast suggest a southerly or south-easterly flow direction. This is in keeping with the south-south-easterly current direction found by Read and Johnson (1967) for the equivalent strata in the Stirling district.

STRATHCLYDE GROUP

A period of localised uplift and erosion terminated deposition of the Inverclyde Group in the district and was followed by a major episode of subaerial volcanicity which produced the lavas and volcaniclastic rocks of the Clyde Plateau Volcanic Formation. Subsequently, uplift and erosion produced the volcanic detritus of the Kirkwood Formation which was deposited locally along the margins of the lava blocks. Elsewhere, and partly contemporaneously, the mainly fluviatile sedimentary rocks of the Lawmuir Formation were deposited. Rocks of the Strathclyde Group occur in a south-west-trending belt about 10 km wide which extends across the district from the north-east corner. They also occupy an extensive area in the south-west, around Paisley.

Clyde Plateau Volcanic Formation

Rocks of the Clyde Plateau Volcanic Formation occur north of Glasgow in a zone about 6 km wide running north-east from Dumbarton forming the high ground of the Kilpatrick Hills and the Campsie Fells. South of the River Clyde they occur in a small area at Old Kilpatrick, in the Beith–Barrhead Hills, and in the Cathkin Braes (Figure 1). Each of these areas forms a discrete, largely fault-bounded lava block with its own stratigraphy and petrographical characteristics (Macdonald, 1975). The lava sequences are all thought to be of similar age but correlation between the blocks is limited. Along the northern and western margins of the Campsie Fells the basal pyroclastic rocks abruptly overlie the Clyde Sandstone Formation but along the south margin the underlying sandstone is progressively cut out in an easterly direction between Spout of Ballagan [572 802] and Campsie Glen [610 800]. In the Kilpatrick Hills the base is transitional in the south-west [426 750] and in the Loch Humphrey Borehole [458 755] where thin sandstones occur in the lowest part of the formation, but farther north the base appears to be disconformable. In the Gleniffer Braes the basal volcaniclastic rocks lie directly on the Kinnesswood Formation, with the Clyde Sandstone and Ballagan formations both absent.

The top of the formation is seen along the eastern flanks of the Kilpatrick Hills where it is overlain by volcanic detritus of the Kirkwood Formation in places. Elsewhere it is overlain by sandstone or conglomerate of the Lawmuir Formation. There is considerable variation in preserved thicknesses in the different areas though the sequences are thought to be penecontemporaneous. The small area of lavas at Old Kilpatrick, which is the top part of the Renfrewshire Hills sequence, is thought to be underlain by up to 1000 m of lavas (Paterson et al., 1990). The sequence in the Campsie Fells, where the top is not preserved, is more than 500 m thick, that in the Kilpatrick Hills greater than 400 m and that in the Beith–Barrhead area less than 300 m. The various sequences are thought to result from one major volcanic episode which took place within a relatively short space of time at the beginning of the Viséan. The rocks of the Clyde Plateau Volcanic Formation are mostly lava with varying proportions of pyroclastic deposits and volcaniclastic rocks in different areas.

Interflow volcaniclastic material forms a very small proportion of the lava pile in distal facies sequences but the proportion increases rapidly around active vents where tephra cores are developed comprising tuffs, agglomerates and breccias, with interbedded highly vesicular lava tongues. In distal facies sequences, the lava sheets, which are mostly massive in character, are separated by weathered zones. Typically, these weathered zones have a crudely stratified layer of scoriaceous volcanic debris overlain by fine-grained red-brown bole-like material. In most cases, this bole-like material has ghost stratification and is thought to be tuff rather than weathered lava top. In proximal facies sequences, fragmental rock with blocks, scoria bombs, spatter and stratified tuffaceous material form a high proportion of the sequence. The lava flows in this facies are highly vesicular and commonly compound, comprising two or more flow units. The texture of the lavas is coarser than that found in distal facies lavas.

A large number of associated vents, plugs and minor intrusions occur in the Campsie Fells, the Kilpatrick Hills and also cutting the older strata lying to the north. Many of these volcanic centres were active contemporaneously and supplied most, if not all, of the lavas in these areas and also perhaps, the topmost part of the Renfrewshire Hills sequence seen at Old Kilpatrick. The source is not known of the lavas in the Beith–Barrhead Hills and in the Cathkin Braes.

16 FOUR LOWER CARBONIFEROUS (DINANTIAN)

Volcanic vents

Many of the volcanic vents in the Kilpatrick Hills and Campsie Fells are demonstrably sources of the local Clyde Plateau Volcanic Formation lavas. Other necks seen cutting the Upper Devonian and Lower Carboniferous strata which lie to the north are mostly small diameter conduits. These contain brecciated lava with varying proportion of disaggregated country rock and are cut by minor intrusions. It is not known whether these conduits ever delivered magma to the surface, but they are petrographically similar to the larger vents and are thought to belong to the same suite. Duncryne Vent [437 859] is unique in that it is the only one in the district seen penetrating Lower Devonian strata at the present level of erosion. This is more than 700 m below the projected base of the Clyde Plateau Volcanic Formation. If the accepted model of an upward-opening funnel is valid, such as is seen in the vent exposed in Craigangowan Quarry [524 769] (Plate 4a), the 300 m diameter at the present surface of erosion would suggest a very large-sized vent at the original level of extrusion. It is petrographically similar to the other Lower Carboniferous vents but contains more fragments of lava rich in ferromagnesian minerals. In some of the larger vents, 'rafts' of sedimentary rock occur for example, Catythirsty Vent [511 813]. Here large 'rafts' of Ballagan Formation mudstones and cementstones have been preserved intact (Whyte, 1968). In other vents, such as Dumgoyne [540 825] (Plate 4b) and Park Hill [536 817],

Plate 4a Coarse agglomerate in funnel-shaped neck of vent cutting fine ash, Craigangowan Quarry, Milngavie (D 1854).

Plate 4b South scarp of Campsie Fells showing trap features formed by lavas of the Clyde Plateau Volcanic Formation. The prominent hills Dumfoyne (left) and Dumgoyne (right) are lower Carboniferous vents (D 2995).

ring-faulting developed and allowed adjacent country rock to subside by as much as 100 m.

Volcanic vents within the lava fields are mostly of large diameter, with their present exposure being at or near the original level of extrusion. These vents are plugged by solidified lava and most of the petrological types occurring in the vents or intrusions can be matched with lavas. Mugearite and hawaiite lavas, however, can only be matched with minor intrusions in the Campsie Fells to the east of the Glasgow district. Some of the plugs are coarser grained than the lavas and a group of small intrusions on Saughen Braes [468 781] show gabbroic textures. This feature is thought to be associated with depths of intrusion in excess of a few hundred metres but not necessarily implying plutonic origin.

Many of the vents occur in groupings which can be related to the NE–SW-trending Dumbarton–Fintry line (Whyte and Macdonald, 1974) or to parallel lines lying to the north-west or south-east. These linear groupings were particularly important during the early phases of eruption of the Campsie lavas where the North Campsie Linear Vent System developed along the Dumbarton–Fintry line. A similar system, referred to as the South Campsie Linear Vent System, developed along the Campsie Fault (Craig and Hall, 1975). Several other linear vent systems are seen in the Campsie Fells (Craig, 1980). The early Kilpatrick Hills lavas appear to have been erupted from small central volcanoes developed along south-westerly projections of these linear systems. The distribution of younger basic lavas between High Craigton [525 770] and Cochno Reservoir [495 760] and on the Kilpatrick Braes suggest that the lavas in these areas were derived from central volcanoes situated near High Craigton [524 768] and Burnbrae Reservoir [474 743] respectively. A line through these centres projected to the north-east intersects the Waterhead Central Volcanic Complex, the source of most of the younger flows on the Campsie Fells.

A less obvious but important NW–SE trend is apparent at the western end of the Kilpatrick Hills where a group of large vents appears to have been a major source area during most of the volcanic episode. A similar trend occurs in Strath Blane at the eastern end of the Kilpatrick Hills where a grouping of major vents was active during the early stages of extrusion (Craig, 1980). These two groupings of vents and their associated tephra cones appear to have been effective barriers which separated the volcanic sequences of the Renfrewshire Hills from the Kilpatrick Hills and of the latter from the Campsie Fells during the greater part of the volcanic episode.

No sources have been identified within the district for the lavas of the eastern Renfrewshire Hills, the Beith–Barrhead Hills, and the Cathkin Braes.

Classification

The great majority of lavas show varying degrees of alteration. The more basic flows are usually fresher, though olivine is almost always pseudomorphed by green or brown bowlingite. Albitisation, chloritisation, carbonation, oxidation and occasionally zeolitisation all occur, leading to considerable difficulty in classifying the rocks either petrographically or petrochemically.

The classification used is essentially the one devised by MacGregor (1928). The more basic rocks mostly contain plagioclase, olivine and pyroxene phenocrysts in varying amounts and on this basis are allocated to six main types (Table 3). Hillhouse, Dalmeny, and Dunsapie types are olivine basalts and the more mafic Craiglockhart type is an ankaramite. Particularly in the Campsie Fells, the feldsparphyric Markle and Jedburgh types include rocks with little or no visible olivine. Analyses indicate that some of these rocks are basaltic hawaiites and hawaiites but they cannot be mapped separately because they occur intimately associated with basalts and are of similar appearance. Following normal field practice, pale grey or pinkish lavas with pronounced platy jointing, caused by flow-alignment of small feldspar crystals, have been classed as mugearites. Trachybasalts with alkali feldspar and plagioclase, a few pink-weathering trachytes, and one rhyolite with restricted distribution, also occur in the Campsie Fells. Intermediate types are common, but composite flows are very rare. Variation within a single flow, particularly in frequency and size of phenocrysts is also very common. Thus a Jedburgh lava may pass along its length into a Markle by increase in number of plagioclase phenocrysts larger than 2 mm.

Western Campsie Fells succession

The western end of the Campsie Fells, which lies in the north-east of the district, is formed by rocks of the Clyde Plateau Volcanic Formation. This area represents about a third of the whole Campsie Block which extends east across most of the northern part of the adjacent Airdrie district (Sheet 31W) and includes the Kilsyth Hills. The Campsie Block is characterised by the presence of a number of source areas, most of which are grouped into linear systems aligned on or parallel to the Dumbarton–Fintry line. The North Campsie Linear Vent System is the only one fully developed in the western Campsie Fells; the others occur mainly in the Airdrie district and are described in Forsyth et al., 1996. Later in the volcanic episode, a major central volcano, the Waterhead Central Volcanic Complex, became established in the Airdrie district and periodically produced considerable quantities of lavas. These source areas were active at various times giving rise to a complex lava stratigraphy and considerable lateral variation over short distances. However, a stratigraphical classification of the volcanic sequence in the Campsie and Kilsyth blocks has been set up by Craig (1980). The succession is split informally into ten subdivisions most of which have several laterally equivalent groups of flows. All ten subdivisions are represented in the Kilsyth Hills but only part of the sequence is present in the western Campsie Fells (Table 4). The distribution of these subdivisions is shown in Figure 6a. The overall sequence preserved in the western Campsie Fells consists of various types of microporphyritic lavas erupted from the linear vent systems interbedded at higher levels with, and overlain by, subdivisions of macroporphyritic lavas derived from the Waterhead Central Volcanic Complex.

Table 3 Classification of basic igneous rocks of Carboniferous and Permian age in the Midland Valley of Scotland.

Basalt type of MacGregor (1928)	Phenocrysts abundant	Phenocrysts sometimes present in lesser amounts	Chemical classification of Macdonald (1975)	Type locality
Macroporphyritic (phenocrysts >2 mm)				
Markle	pl	± ol, Fe	pl ± ol ± Fe-phyric basalts, basaltic hawaiites or hawaiites	Markle Quarry, East Lothian (flow)
Dunsapie	pl + ol + cpx	± Fe	ol + cpx + pl ± Fe-phyric basaltic hawaiites or ol + cpx + pl - phyric basalts	Dunsapie Hill, Edinburgh (vent intrusion)
Craiglockhart	ol + cpx		Ankaramite	Craiglockhart Hill, Edinburgh (flow)
Microporphyritic (phenocrysts < 2 mm)				
Jedburgh	pl	± ol, Fe	pl ± ol ± Fe-phyric basaltic hawaiites and in some cases basalt	Little Caldon, Stirlingshire (plug). Also in Jedburgh area
Dalmeny	ol	± cpx, pl	ol ± cpx - phyric basalt	Dalmeny Church, West Lothian (flow)
Hillhouse	± ol + cpx		ol ± cpx-phyric (rarely basanite)	Hillhouse Quarry, West Lothian (sill)

pl = plagioclase, ol = olivine, cpx = clinopyroxene, Fe = iron-titanium oxides

Campsie Lavas: Lower (3A) and Upper (5A) North Campsie Lavas; Lower (3B) and Upper (5B) South Campsie Lavas
The Campsie Lavas are the oldest flows seen in the western Campsie Fells. They are typical Jedburgh-type lavas derived mainly from the North and South Campsie linear vent systems, although the Gonachan Glen and Dungoil linear vent systems were active in the north during the later stages of extrusion. They vary in composition from basalt to hawaiite and have a considerable range of modal compositions, particularly with respect to their clinopyroxene content. Many of the flows derived from the south contain significantly more clinopyroxene than those derived from the north. Though it is possible to assign most of the flows in the north and south of the area to their appropriate sources, in the intervening area, where interdigitation is thought to have taken place, it has not been found practical to separate them. Upper and Lower lavas can only be separated with certainty where the distinctive Craigentimpin Lavas intervene. Within the district, this only happens in the extreme south-east of the Campsie Block. The maximum overall thickness for the Campsie Lavas of about 260 m occurs north of Gonachan Glen but this thins to 200 m in the north-west and only 100 m in the south. This variation is attributed to more prolonged and vigorous effusion in the north than in the south. The Lower North Campsie Lavas are well exposed at Black Craig [558 812] where there is a sequence of nine flows with well-developed interflow horizons. The lowest lava is more basic than the overlying flows and is assigned to the Lower South Campsie Lavas. The North Campsie Lavas are about 90 m thick here and display crude columnar jointing.

Petrologically they consist of abundant microporphyritic plagioclase of a rather sodic andesine composition, moderate amounts of microporphyritic olivine and very sparse augite.

The Lower South Campsie Lavas are well exposed in the Forking Burn [653 790] where the sequence is 80 m thick and is composed of ten proximal-facies Jedburgh flows showing toe structures enveloped in slaggy lava, typical of pahoehoe flows. Petrologically the lavas consist of microporphyritic plagioclase of sodic labradorite and a generally quite high content of mafic minerals which have been extensively replaced by carbonate. There are no good sections through the Upper North Campsie Lavas but in the tributaries of the Gonachan [602 838] and Clachie [612 840] burns, faulted and partly drift obscured sequences occur. These are typical proximal facies sequences with thin carbonated lavas and considerable amounts of scoriaceous interflow agglomerates and tuffs. Just to the east in the Airdrie district, in the Muir Toll Burn [628 828], the Upper North Campsie Lavas are about 40 m thick and consist of a series of trachybasalt flows.

The Upper South Campsie Lavas only occur in the southern part of the Campsie Block where they interdigitate with lavas derived from the north. They are exposed in the Aldessan [608 805] and Alvain [617 806] burns where they are distinguished from those derived from the north by their more basic composition.

Craigentimpin Lavas (4A) The Craigentimpin Lavas are distinctive macroporphyritic lavas of Markle type with large platey phenocrysts of calcic labradorite, repre-

Table 4 Stratigraphy of the Clyde Plateau Volcanic Formation in the western Campsie Fells (after Craig, 1980).

		Lava types	Thickness (m)	Derivation
Knowehead Lavas	10	Mainly B^M and B^{Du} with B^J, W^M and R	50+, top faulted	Local centres
Holehead Lavas	9	Dominantly B^M	100+, top eroded	Waterhead Central Volcanic Complex
Fin Glen Lavas	8	Mainly B^J and WB, with persistent minor T	110	Local centres and North Campsie Linear Vent System
Upper North Campsie Lavas	5A	BW and B^J	215	North Campsie, Gonachan Glen and Dungoil linear vent systems
Upper South Campsie Lavas	5B	BW and B^J	60	South Campsie Linear Vent System
Craigentimpin Lavas	4	B^M with large feldspar phenocrysts	0–30, absent in western part of area	Waterhead Central Volcanic Complex
Lower North Campsie Lavas	3A	BW and B^J	70	North Campsie Linear Vent System
Lower South Campsie Lavas	3B	B^J and BW	135	South Campsie Linear Vent System

B^{Du} — Dunsapie, B^J — Jedburgh, B^M — Markle, BW — Basalt/trachybasalt, R — Rhyolite
T — Trachyte, W^M — Mugearite

sented in the district by a single flow. It is well exposed in Craigentimpin Quarry [616 802] and the Alvain Burn [618 805]. The lava thins out farther west and the Upper Campsie Lavas lie directly on Lower Campsie Lavas at the western end of the Campsie Fells. Orientation of the feldspar phenocrysts suggests that these flows were derived from the Waterhead Central Volcanic Complex (Whyte and McDonald, 1974).

Fin Glen Lavas (8A) The Upper North and South Campsie lavas are overlain by the Fin Glen Lavas. The lavas which make up this member are microporphyritic, more felsic than the underlying flows, and characterised by mugearitic and trachybasaltic types. A thin but widespread development of trachytes occurs at the base of the Fin Glen Lavas. The trachytes bear a close petrological resemblance to the phonolytic trachyte intrusion which occurs in the North Campsie Linear Vent System near Fintry [614 863] and may be derived from it. A good section is seen in the Aldessan Burn [608 807].

Holehead Lavas (9A) The Fin Glen Lavas are succeeded by the Holehead Lavas, the youngest lavas preserved in the western Campsie Fells. The Holehead Lavas are macroporphyritic olivine-basalts of Markle type which are thought to have been derived from the Waterhead Central Volcanic complex.

Knowehead Lavas (10A) South of the Campsie Fault, the Knowehead Lavas, a fault-bounded sequence of flows, occurs which cannot be correlated with any part of the Campsie Fells succession. They are more akin to the Cochno Lavas of the Kilpatrick Hills succession and are thought to be younger than anything preserved in the Campsie Fells. These lavas include olivine-basalts of Dunsapie type and also the only rhyolite flow [602 796] in the whole district. Near Clachan of Campsie [610 797] a thick development of tephra deposits suggests a local source for at least some of these lavas.

KILPATRICK HILLS SUCCESSION

Rocks of the Clyde Plateau Volcanic Formation form the Kilpatrick Hills which extend north-east from Bowling to Strathblane. In this area the succession has been divided into members (Table 5) and, since the general dip is to the south-east, the various members occur in roughly NE–SW-trending belts becoming progressively younger to the south-east (Figure 6b). Unlike neighbouring areas, the lavas in the Kilpatrick Hills are the products of two distinct pulses of extrusive activity separated by a quiescent phase. During the latter, erosion took place followed by the deposition of volcanic detritus. In the west, the volcanic detritus is overlain by a thin sequence of clastic and organic rocks, followed by tuff. The first pulse of extrusive activity is characterised by contemporaneous eruption from several centres of different types of lava giving rise to a complex sequence of neighbouring and overlapping units (Hamilton, 1956). All the lava members that were erupted during the first episode thin to-

Figure 6
Stratigraphical divisions of the Clyde Plateau Volcanic Formation in different lava blocks.

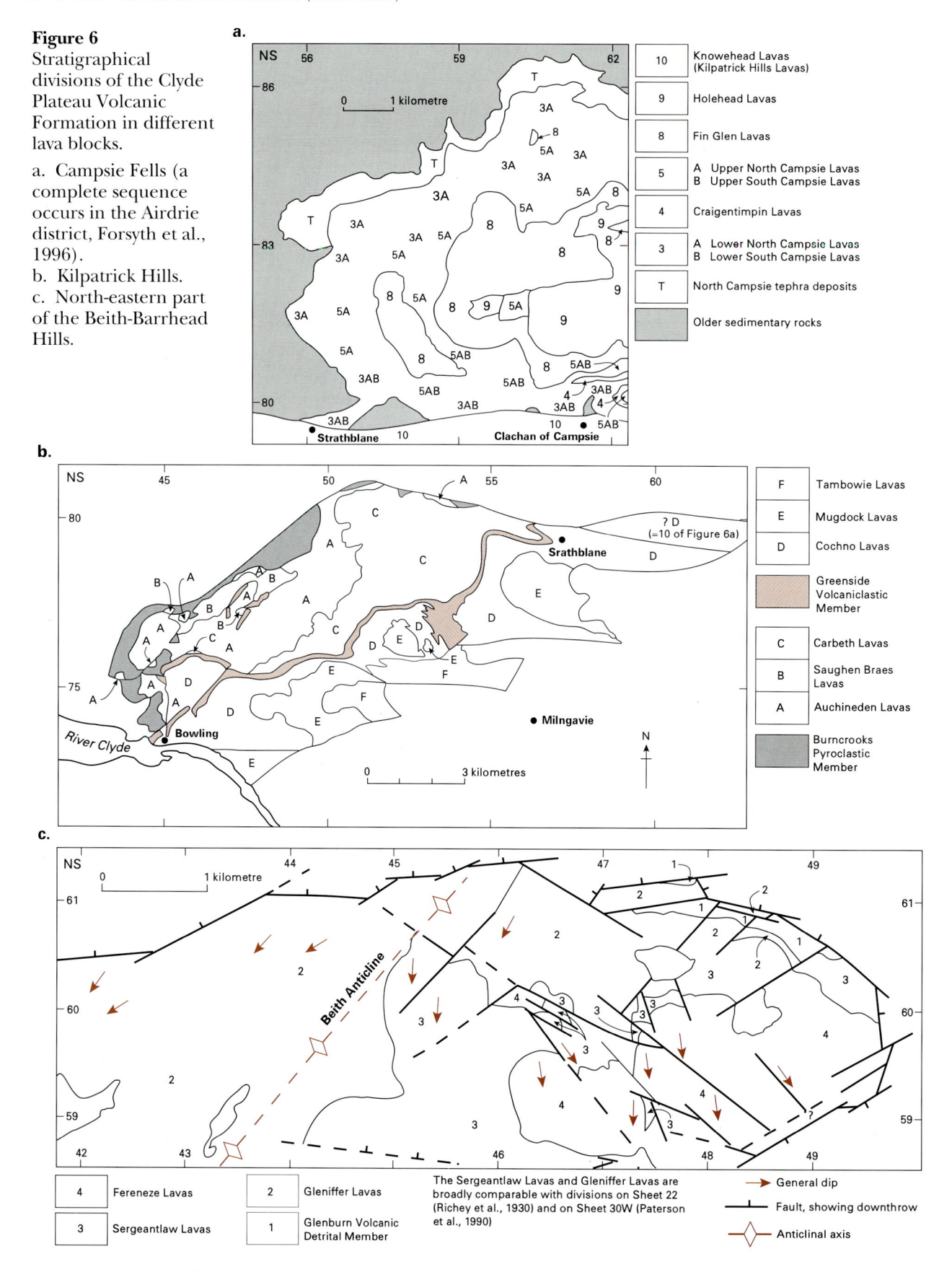

a. Campsie Fells (a complete sequence occurs in the Airdrie district, Forsyth et al., 1996).
b. Kilpatrick Hills.
c. North-eastern part of the Beith-Barrhead Hills.

Table 5 Stratigraphy of the Clyde Plateau Volcanic Formation in the Kilpatrick Hills.

	Lava types	Thickness (m)	Derivation
Tambowie Lavas	Alternations of B^D and W^M	maximum 30, absent in NE	Distal facies derived from south
Mugdock Lavas	Dominantly B^M with W^M, B^{Ck}	100–200, thins to SW	?Waterhead Central Volcanic Complex
Cochno Lavas	Dominantly B^M, B^J, with B^D, B^{Ck}, B^{Du}, R	100–250, thins to NE, but thicker adjacent to Campsie Fault	Local centres at Bowling, Burnbrae, Black Loch, Craigton, and Clachan of Campsie for basic lavas and rhyolite; rest partly from Bowling area
Greenside Volcaniclastic Member	Volcaniclastic and lacustrine sediment, ash, tuff, and agglomerate	Mostly 5–10, 40 at Loch Humphrey and 55 at Craigton	Burnbrae, Black Loch, Craigton, Clachan of Campsie
Carbeth Lavas	Dominantly B^M with minor W^M and B^{Du}	0–150, thins out to SW partly due to erosion; thins to >80 in NE	Some local, rest unknown
Saughen Braes Lavas	B^M	0–40, only present in north-central area	Local, derived from north ?Knockupple
Auchineden Lavas	Mainly B^J, some W^M rare B^{Ck}, B^D, B^{Du} and B^M	0–100, very variable, thins out to NE, and SW	Local centres, such as Burncrooks, Kilmannon, Doughnot Hill, and ?Brown Hill
Burncrooks Pyroclastic Member	Volcaniclastic and lacustrine sediment, ash, tuff and agglomerate	0–80, very variable, thins to NE and south	Various tephra cones: Burncrooks, Brown Hill, Catythirsty, etc.

B^{Ck} — Craiglockhart, B^D — Dalmeny, B^{Du} — Dunsapie, B^J — Jedburgh, B^M — Markle, R — Rhyolite, W^M — Mugearite

wards the partly exhumed tephra cone of Brown Hill [442 762]. The second pulse is marked by the much more widespread development of each of the lava members. The lavas are predominantly olivine-basalt in composition but ankaramites, hawaiites and mugearites are also present.

Burncrooks Pyroclastic Member Volcanicity started in the Kilpatrick Hills with the eruption of pyroclastic material from several vents. The pyroclastic material was deposited partly as tephra cones which merged in some places. Elsewhere the intervening ground was covered by airfall tuff and reworked volcaniclastic sediment derived from the tephra cones. These rocks mostly crop out along the north-western and western margins of the lava block. At Burncrooks [482 798], where one centre was located, the pyroclastic rocks are about 70 m thick and are partly waterlain at higher levels. They become thinner to the north-east, as at Catythirsty [515 809]. South-west from Burncrooks they are brought to the surface by faulting around Fynloch [460 772] and Lily Loch [473 780] where they are only 10 to 20 m thick and consist mainly of ash and volcaniclastic rocks. At the southern end of Lang Craigs [436 757] they are about 80 m thick around the partly exhumed tephra cone of Brown Hill and consist mostly of airfall tuff and volcaniclastic rocks. Further south, at Rigangower [438 752], the top 2 m consist of lacustrine shales containing fish scales which overlie a discontinuous coal seam up to 0.5 m thick. Roots in the seatrock beneath the coal are partly replaced by the zeolite, heulandite. This is the only area within the Kilpatrick Hills where such sedimentary rocks are known to occur. The lateral extent of the Burncrooks Pyroclastic Member beneath the younger lavas is unknown but 70 m of pyroclastic rocks, mostly airfall tuff, were proved in the Loch Humphrey Borehole [4592 7555].

Auchineden Lavas The eruption of the Burncrooks Pyroclastic Member was followed by the almost simultaneous eruption of dominantly microporphyritic lavas of Jedburgh type, the Auchineden Lavas, and macroporphyritic lavas of Markle type, the Saughen Braes Lavas. The former are well exposed on the west flank of Auchineden Hill [493 805] where eight or nine Jedburgh type flows, totalling 100 m in thickness, gradually thin to the south and interdigitate with four or five trachybasaltic flows which thin to the north. The latter show typical proximal facies characteristics near Kilmannon Resevoir [495 784] where the lavas are interbedded with cone facies tuffs. The Auchineden Lavas gradually thin out towards the north-east and at Arlehaven [535 802] are absent. West of Burncrooks [468 785] they interdigitate with and overlie the Saughen Braes Lavas and further west on Lang Craigs [430 760] competely replace them. On the Lang Craigs, where the sequence is about 80 m thick, the lowest widespread flow is a Craiglockhart lava of ankaramitic composition. In the north-east of this area [443 772] it is underlain by one flow of Markle-type basalt belonging to the Saughen Braes Lavas and elsewhere by a thin discontinuous flow of Dalmeny type. This is the only area where flows of a more basic composition are found at the base of the sequence. They are possibly derived from one of the small vents such as Knockshannoch [434 798] lying to the north-west where similar material is preserved in a small neck. The remainder of the lavas on Lang Craigs are thought to have been derived from the Doughnot Hill Vent [447 776] or possibly the Brown Hill Vent [443 765]. The extent of the Auchineden Hill Lavas below younger lavas is not known, but they were 45 m thick in Loch Humphrey Borehole [4592 7555].

Saughen Braes Lavas These Markle-type lavas were erupted simultaneously with the early Auchineden Lavas but only occur on Saughen Braes [462 783]. They comprise about five flows of Markle-type lava totalling 40 m in thickness. To the north-east and south-west they abut against the Auchineden Lavas and to the south they thin out near Fynloch [458 770]. They appear to be derived from the north, possibly from the Knockupple Vent [458 793].

Carbeth Lavas The succeeding, geographically widespread Carbeth Lavas are mostly macroporphyritic lavas of Markle type which extend from Strath Blane in the north-east to Brown Hill in the south-west. These lavas also include some Jedburgh and Dunsapie flows and a few mugearites near the base. This member is thickest in the central part of the Kilpatrick Hills where it attains a maximum of about 150 m. The lavas are well exposed around Carbeth [527 796] where a few Dunsapie flows are intercalated and in the Birny Hills [495 770] where scarp features are well developed. West of Loch Humphrey [455 760] the member is reduced to one Dunsapie flow which is only preserved in isolated patches below the unconformity that occurs at the base of the Greenside Volcaniclastic Member. The source of these lavas is not known as proximal facies characteristics are nowhere developed.

Greenside Volcaniclastic Member This member consists of sedimentary and volcaniclastic rocks that can be traced across the Kilpatrick Hills from Strathblane to Bowling. These rocks mark a major break in the lava succession. They were formerly thought to be a faulted repetition of the basal tuffs seen at Burncrooks (Burncrooks Pyroclastic Member) and consequently to occur immediately below the main sequence of lavas (Scott et al., 1984). However, detailed mapping suggested that the faulting was relatively minor and that the succession seen at Greenside was much higher in the sequence than was previously thought. The Loch Humphrey Borehole [4592 7555] penetrated the lowest 20 m of the Greenside Volcaniclastic Member overlying 140 m of lavas and tephra which are correlated with the Carbeth Lavas, the Auchineden Lavas and the Burncrooks Pyroclastic Member. The thin coal seams, plant-bearing beds (Currie, 1866; Young, 1873b) and mudstone with fish (Young, 1873a) which occur near the top of the Greenside Volcaniclastic Member at Auchentorlie [441 741], Glenarbuck [452 747] and Loch Humphrey Burn [468 755] have been known for a considerable time. These organic and clastic sedimentary rocks do not occur east of Greenside [475 755] where they are represented by tephra deposits, as at Craigton [525 770], and by tuff and volcaniclastic sedimentary rocks, in the Strathblane area. This member is variable in thickness, but is generally mostly 5 to 10 m. However, where it merges into contemporary tephra cones, as at Clachan of Campsie and Craigton, it is more than 50 m thick. Also at Greenside, where greater subsidence allowed lacustrine conditions to develop, 40 m of sedimentary rocks accumulated.

Cochno Lavas Eruption of lavas restarted in the Kilpatrick Hills with the extrusion of the Cochno Lavas. This member includes a wide range of lava types but is characterised by the presence of more basic flows including Dalmeny, Craiglockhart and Dunsapie types, particularly near the base. In the south-west, around the Bowling Centre [445 738] which appears to have been producing localised Markle and Jedburgh lavas at this time, the only basic flow is a widespread Dalmeny lava at the base of the member. Further east, Craiglockhart and Dunsapie types appear at higher levels and are spatially related to vents at Burnside [474 741], Black Loch [500 765] and Craigton [525 770]. The vents all have plugs of these rock types. Individual lava flows of this member show marked variation in thickness over short distances, indicative of extrusion onto irregular surfaces. This is evident on the Kilpatrick Braes [470 740] and between Cochno Loch [510 765] and Craigton. On the Kilpatrick Braes, the basic flows appear to occur in channels radiating out from a centre near Burnside. In contrast, near Craigallian Loch [537 785], some flows can be traced over 2–3 km with only minor changes in thickness. In this area three of the flows show good columnar jointing where the lavas cooled in ponds. The Knowehead Lavas of the Western Campsie Fells (Table 4) which contain the only rhyolitic flow known to occur in the district, are correlated with the Cochno Lavas. The Knowehead Lavas appear to be locally derived from a centre near Clachan of Campsie [610 797]. The Bowling, Burnside, Blackloch, Craigton and Clachan of Campsie vents all lie in a narrow zone parallel to the Dumbarton–Fintry Line which, when projected to the north-east, intersects the Waterhead Central Volcanic Complex.

Mugdock Lavas   The Cochno Lavas are succeeded by the Mugdock Lavas which are predominantly of Markle type. They are well exposed around Mugdock Loch [554 773] and at Devil's Craig Dam [560 782]. Near the latter, a few flows have frequent feldspar phenocrysts of about 5 mm and occasional ones up to 2 cm in length. Two or three of the highest flows have reddened ferromagnesian phenocrysts in otherwise unoxidised lava. Near Craigend Castle [548 777] a thin Craiglockhart flow occurs and in the western part of the Kilpatrick Hills, intercalations of mugearite are present. The Mugdock Lavas at Old Kilpatrick [465 730] are continuous with a similar sequence seen south of the River Clyde (upper part of the Strathgryfe Lavas; Paterson et al., 1990). This member is about 200 m thick in the north-east and thins to about 100 m in the south-west; it may have been derived from the Waterhead Central Volcanic Complex.

Tambowie Lavas   The Tambowie Lavas are the youngest flows in the Kilpatrick Hills. They are only present in the southern part of the area where they are about 30 m thick. They are absent in the north-east near Muirhouse [563 780] where the overlying sedimentary rocks of the Lawmuir Formation lie directly on Mugdock Lavas. They are not seen in the south-west where they may be cut out by faulting. The Tambowie Lavas mainly consist of Dalmeny basalts interbedded with mugearites. Near Edinbarnet [503 745] distal facies characteristics such as lava toes and thin reddened lavas with interbedded boles are commonly developed. The highest flow seen in the Douglas Muir area [525 751] is of Dunsapie type with an unusual 'cumulate' texture. The source of the Tambowie Lavas is unknown but they appear to have been derived from the south or south-east.

RENFREWSHIRE HILLS SUCCESSION

The small areas of poorly exposed lavas which occur at Old Kilpatrick [450 720] and south of Bishopton [424 694] are the easterly fringes of the Renfrewshire Hills sequence seen to the west in the Greenock district. The Markle-type lavas at Old Kilpatrick are the highest part of the 750 m-thick Strathgryfe Lavas (Paterson et al., 1990). They are continuous with the Mugdock Lavas of the Kilpatrick Hills. The Craiglockhart and Dunsapie flows which occur at Langbank to the west of the Glasgow district, may correlate with the Cochno Lavas. The small area of lavas south of Bishopton is part of the Marshall Moor Lavas (Paterson et al., 1990) and may be equivalent to the Tambowie Lavas of the Kilpatrick Hills sequence.

BEITH–BARRHEAD HILLS SUCCESSION

The Beith–Barrhead Hills cover a total area of 16 km by 6 km, extending to the south-west and west of the Glasgow district into sheets 22E, 22W and 30W. The lavas are almost all of olivine-basalt composition (Table 6) and

Table 6   Divisions of the Beith–Barrhead lava succession.

New names used by Paterson et al., 1990	Lava types	Names used by Richey et al., 1930
(Beith Lavas)		(Upper Group) (Sheet 22)
Fereneze Lavas	B^J, B^M, B^D, B^{Du}	Lower Group (c)
Sergeantlaw Lavas	B^D, B^{Du}	Lower Group (b)
Gleniffer Lavas	B^M	Lower Group (a)
Glenburn Volcanic Detrital Member		—

B^D — Dalmeny, B^{Du} — Dunsapie, B^J — Jedburgh, B^M — Markle

have a possible total thickness of up to 300 m. They occur in a fault-bounded, broad, shallow NE-trending anticline which plunges to the south-west (Figure 6c). In north Ayrshire, Richey (*in* Richey et al., 1930, p.70) recognised four divisions, based upon outcrops around the closure of the anticline and on its north-western limb, where a similar succession may be traced north-eastwards into the Greenock district (Paterson et al., 1990). On the south-eastern limb, much of which is included in the Glasgow district, the correlation is less certain.

Gleniffer Lavas   Owing to the plunge of the anticline, the lowest lavas, the Gleniffer Lavas which correspond to Richey's Lower Group (a), occur in the north of the outcrop along the prominent fault scarp which extends from Gleniffer Braes [450 607] to Brownside Braes [485 605]. The base of the main lava sequence is well exposed at Linn Well [475 605] in the Glen Park, Paisley, an undercut waterfall formed by a massive flow of Jedburgh lava which rests upon well-bedded, purple-brown, volcaniclastic sandstones and siltstones of the Glenburn Volcanic Detrital Member. These sedimentary rocks are exposed at intervals along the base of the lava scarp, eastwards for 1.6 km to Harelaw Burn. In the BGS Glenburn Borehole [478 607], which was sited on a downfaulted block just to the north of the main scarp, lavas with interbedded agglomerate and volcaniclastic sedimentary rocks rest directly upon sandstones of the Kinnesswood Formation. Most of the Gleniffer Lavas exposed in the core of the anticline, from Whittliemuir Midton Loch [420 590] to Gleniffer Braes and from thence eastwards to Glenburn Reservoir [475 600], are olivine-basalts of a very distinctive Markle type, characterised by abundant, platy phenocrysts of plagioclase up to 2.5 cm long and microphenocrysts (<2 mm) of olivine.

Sergeantlaw Lavas   The Sergeantlaw Lavas consist of a sequence of more mafic olivine-basalts of Dunsapie type with some of Dalmeny type, which crop out over a wide area on the south-eastern limb of the anticline, from Sergeant Law to Greenfieldmuir [456 583] and in the

Lochliboside Hills [450 575] in the Kilmarnock district (Sheet 22E). These may correlate broadly with Richey's Lower Group (b). Similar lavas occur to the north of Harelaw Reservoir [485 590] where they appear to cut out or overlap most of the Gleniffer Lavas and are thus separated from the basal volcaniclastic sedimentary rocks by only one or two flows of 'big-feldspar' Markle basalt. To the south-east of Harelaw Burn, around Boylestone Quarry, Barrhead, mafic lavas appear to rest directly upon the basal volcaniclastic rocks. A thick flow of Dalmeny basalt in Boylestone Quarry [495 597] is renowned for its hydrothermal veins and cavities which contain a wide variety of minerals including calcite, analcime, prehnite, heulandite, bowlingite, thompsonite and natrolite. Native copper, cuprite and malachite have also been recorded from joint surfaces and as disseminations. A thin layer of well-bedded volcaniclastic sedimentary rock lies on top of and infills erosive channels cut into the Dalmeny flow. The channels are well seen in the quarry face, beneath the overlying flow of Jedburgh lava.

Fereneze Lavas The youngest lavas in this succession are the Fereneze Lavas, which occur in the east of the hills between Capellie Farm [465 583] and the top of Boylestone Quarry. They cannot be correlated directly with Richey's groups but may be equivalent to his Lower Group (c). Markle basalts and Jedburgh lavas, some of hawaiitic composition, are overlain by a sequence of Dalmeny, 'big-feldspar' Markle and Dunsapie basalts in the Fereneze Hills. As is the case with the Sergeantlaw Lavas, the Fereneze Lavas appears to cut out or overlap most of the preceding lavas. It also seems likely that the Gleniffer Lavas of the Beith–Barrhead Hills die out progressively towards the east, probably against a rising topography. This may have been due, at least in part, to previous or contemporaneous movement on NE-trending faults, parallel to the major fractures of the Paisley Ruck and Dusk Water Fault. Such movements must be inferred for example south-east of Glenburn Reservoir and along Harelaw Burn where marked changes in lava stratigraphy occur in close proximity to the positions of post-lava NE-trending faults.

Cathkin Braes

In the south-east of the district a small area of poorly exposed lavas of the Clyde Plateau Volcanic Formation occurs in the Cathkin Braes. The base of the lavas is faulted so that only a minimum thickness of 60 m can be estimated for the sequence. The succession consists of a lower assemblage, more than 30 m thick, of microporphyritic olivine-basalts mainly of Dalmeny type, and an upper assemblage about 30 m thick of macroporphyritic olivine-basalts of Dunsapie type. Thin intercalations of Jedburgh and Markle lavas and agglomerate occur in the lower assemblage; a thin agglomerate occurs in the upper.

Regional correlation

The correlation (Figure 7) of the lava successions in the western Campsie Fells with those of the Kilsyth, Fintry, Gargunnock and Touch hills is discussed in Forsyth et al. (1996). The lavas of the western Campsie Fells and the Kilpatrick Hills are juxtaposed along the Campsie Fault and separated by only 2–3 km along Strath Blane. A general correlation has been suggested (Grabham, in Macgregor et al., 1925) of the Jedburgh basalts on Auchineden Hill (the Auchineden Lavas) with the upper part of the sequence of Jedburgh-type lavas widely exposed on the western Campsie Fells (?North Campsie Lavas), and of the various macroporphyritic lavas seen in the Kilpatrick Hills (Cochno and Mugdock lavas) with the upper group of the western Campsie Fells, north of the Campsie Fault (Holehead Lavas). Whyte and MacDonald (1974) suggested that the Markle basalts south of the Campsie Fault (Cochno and Mugdock lavas) may be equivalent to the upper flows of the Campsies (Holehead Lavas) north of the fault, some 300 m higher topographically, but point out that appreciable erosion prior to deposition of the overlying sedimentary rocks would render correlation uncertain. Craig (1980) classified the Knowehead Lavas, which lie to the south of the Campsie Fault, as part of the youngest assemblage of the Campsie succession and implied that they have been eroded off the western Campsie Fells north of the Campsie Fault. At the eastern end of the Kilsyth Hills, the Garvald Lavas which are thought to be a similar age, are overlain by the Kirkwood Formation.

Despite the close proximity of the sequences in the western Campsie Fells and the Kilpatrick Hills, systematic mapping has shown that there is no more than a superficial similarity between the two: both have microporphyritic lavas of Jedburgh type succeeded by macroporphyritic lavas of dominantly Markle type. Petrographically, however, the lavas of the Kilpatrick Hills succession are less felsic and have more abundant clinopyroxene than those of the Western Campsie Fells succession. In addition, the Kilpartick Hills sequence is interspersed throughout with basic lavas of Dalmeny, Dunsapie and Craiglockhart types whereas they are absent from the western Campsie Fells north of the Campsie Fault. Many of these basic flows are thought to be of local origin and may never have reached the western Campsie Fells but those in the Cochno Lavas (Knowehead Lavas) are well developed just south of the fault and would be expected in the western Campsie Fells sequence. The most likely explanation seems to be that the Cochno Lavas are younger than any flows preserved on the western Campsie Fells and correlate with the Garvald Lavas at the eastern end of the Kilsyth Hills, as inferred by Craig (1980). In addition, all the members of the Kilpartick Hills sequence below the Cochno lavas thin towards or are absent in the Strath Blane area, suggesting that they were geographically isolated from the western Campsie Fells sequence. The direction of thinning indicates deposition on a palaeoslope which rose towards Strath Blane. In the Campsie Fells sequence, the Craigentimpin Lavas thin out to the west also suggesting high ground in the Strath Blane area. It is known that the major vents of Dumgoyne, Dumfoyne and Park Hill were active during the early stages of the volcanic episode (Craig, 1980) and are presumed to have developed tephra cones. It seems probable that during these early stages the high ground

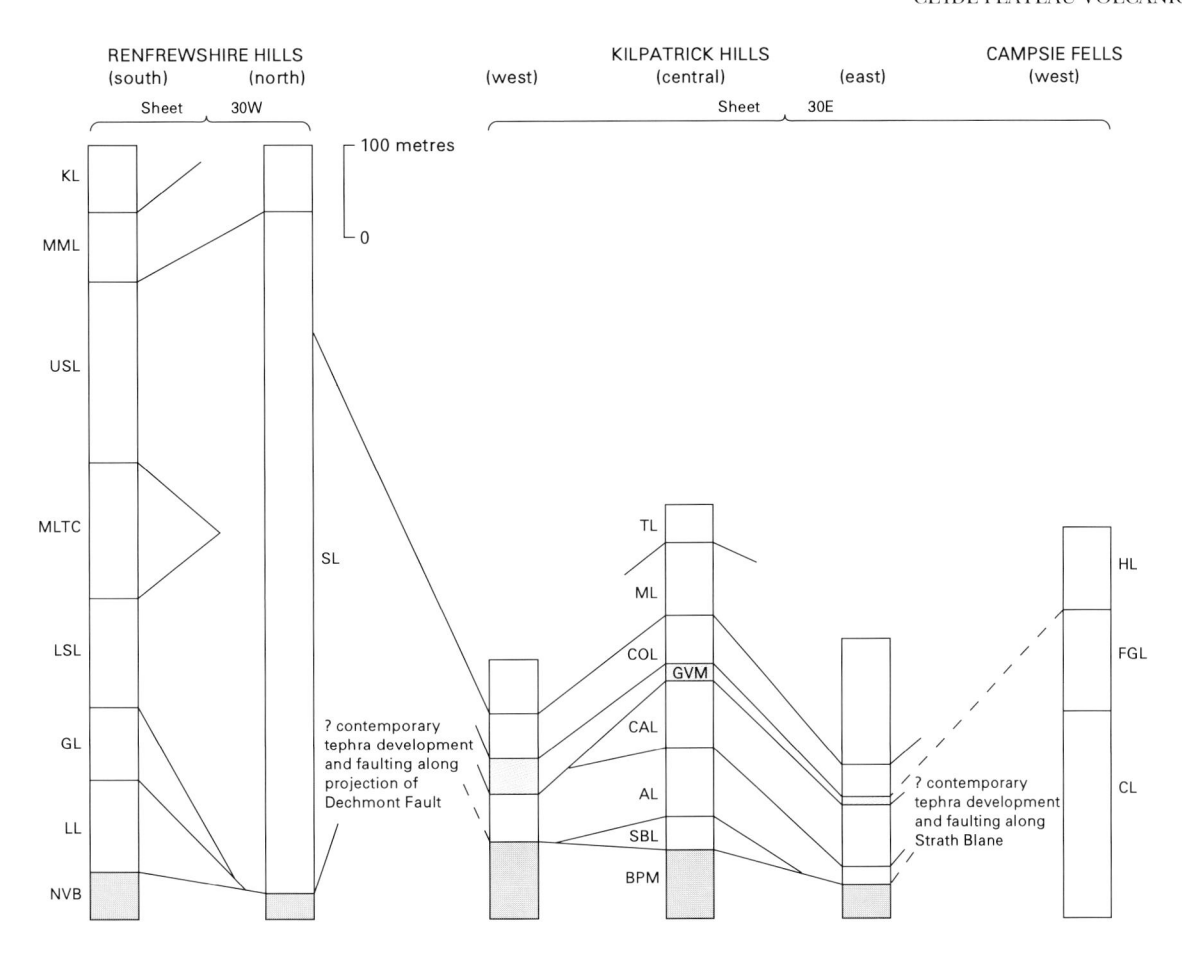

Figure 7 Composite sections showing the correlation of the Clyde Plateau Volcanic Formation in the Kilpatrick Hills and with adjacent areas. Sequence in the Renfrewshire Hills after Stephenson, *in* Paterson et al., 1990; in the Campsie Fells, after Craig, 1980.

Abbreviations: AL — Auchineden Lavas, BPM — Burncrooks Pyroclastic Member, CL — Campsie Lavas (Upper North and South, Lower North and South), CAL — Carbeth Lavas, COL — Cochno Lavas, FGL — Fin Glen Lavas, GL — Greeto Lavas, GVM — Greenside Volcaniclastic Member, HL — Holehead Lavas, KL — Kilbarchan Lavas, LL — Largs Lavas, ML — Mugdock Lavas, MML — Marshall Moor Lavas, MLTC — Misty Law Trachytic Centre, NVB — Noddsdale Volcaniclastic Member, SBL — Saughen Braes Lavas, SL — Strathgryfe Lavas (USL — Upper, LSL — Lower), TL — Tambowie Lavas.

that was an effective barrier between the two areas, was formed by these tephra cones. Above the Cochno Lavas, the dominantly Markle flows of the Mugdock Lavas are thickest in the north-east of the Kilpatrick Hills and thin towards the south-west. This member may have been derived from the Waterhead Central Volcanic Complex and originally may have been present on the western Campsie Fells. If this is the case, the barrier formed by the tephra cones became ineffective shortly after these vents ceased activity. The highest lavas of the Kilpatrick Hills sequence (the Tambowie Lavas) are not present in the north-east end of the Kilpatrick Hills and would not be expected to reach the Campsie area.

At the western end of the Kilpatrick Hills a similar situation occurs where only the higher parts of the sequence can be matched with the succession in the Renfrewshire Hills. The Tambowie Lavas of the Kilpatrick Hills sequence, characterised by Dalmeny and mugearite flows, may be equivalent to the basic lavas south of Bishopton which are part of the Marshall Moor Lavas (Paterson et al., 1990). The underlying Markle lavas occurring at Old Kilpatrick and the Mugdock Lavas of the Kilpatrick Hills sequence extend south of the River Clyde where they form the upper part of the Strathgryfe Lavas of Paterson et al. (1990). The Craiglockhart flows that occur at Slateford [424 725] and to the west of the Glasgow district near Langbank, are basic developments in the Strathgryfe Lavas. They occur at about 550 m and 750 m above the base of the unit respectively. These Craiglockart flows are thought to be equivalent to the basic flows which occur within and at the base of the Cochno Lavas. Below this level the two sequences are quite different. South of the Clyde, the sequence is made up of more than 500 m of Strathgryfe Lavas which are dominantly mugearite and Markle flows of hawaiitic composition. North of the River Clyde, the equivalent sequence consists of some 40 m of Auchineden Lavas comprising a basal Craiglockhart lava and a few thin mainly Jedburgh-type flows, overlying about 30 m of the Burncrooks Pyroclastic Member. As far as the Kilpatrick Hills sequence is concerned, the lowest three members all thin to the south-west. The Carbeth Lavas and the Saughen Braes Lavas do not reach the area and the Auchineden Lavas are reduced in places to one or two thin flows. In the Kilpatrick Hills, deposition of these lava members appears to be controlled by a land surface which rose in a westerly direction. This palaeoslope is likely to represent the sides of the tephra cones formed by the NW–SE-aligned group of vents which occur at the western end of the area. It appears that these tephra cones became buried during the later stages of the volcanic episode, allowing lava flows to spread over the whole area.

GEOCHEMISTRY

The lavas are all members of a mildly alkaline, alkali basalt series intermediate between the sodic alkali basalt series of the British Tertiary Provence and the potassic alkali basalt series of Tristan da Cunha. The range is ankaramite-basalt-hawaiite-mugearite-benmoreite-trachyte-rhyolite with the end members being less common and benmoreite unrepresented in the district (Figure 8). The available analyses show that a few of the more basic rocks are nepheline-normative but most are hypersthene-normative. All the less basic, more fractionated rocks are hypersthene-quartz-normative.

Studies by Macdonald (1975), Macdonald et al. (1977), Craig (1980), MacDonald and Whyte (1981), and Smedley (1986; 1988) all suggested that the alkali basalt series present in the Campsie Block (Figure 8) is derived by fusion of sublithospheric upper mantle material. Such magmatism is typical of that which occurs in continental rift environments throughout the world. The evolved condition of the lava flows indicates fractionation of clinopyroxene at high pressures in excess of 10 kilobars. This would effect the transition from basalt to basaltic hawaiite. Subsequent lower-pressure fractionation at higher levels, dominated by crystallisation of olivine, plagioclase and metallic oxide, would give rise to the other members of the series. Extraction of substantial amounts of clinopyroxene from the parental magma is supported by the relative paucity of clinopyroxene in many of the lavas in the Campsie Block (Craig, 1980). The variation seen in the lava types erupted from the Waterhead Central Volcanic Complex may have been

Figure 8 Compositional range of lavas within the Glasgow district.

The classification is based on CIPW norm calculations, after Coombs and Wilkinson (1969) and Macdonald (1975). Beith–Barrhead Hills and Kilpatrick Hills, unpublished BGS analyses by J G Fitton, University of Edinburgh; Campsie Fells after Craig (1980).

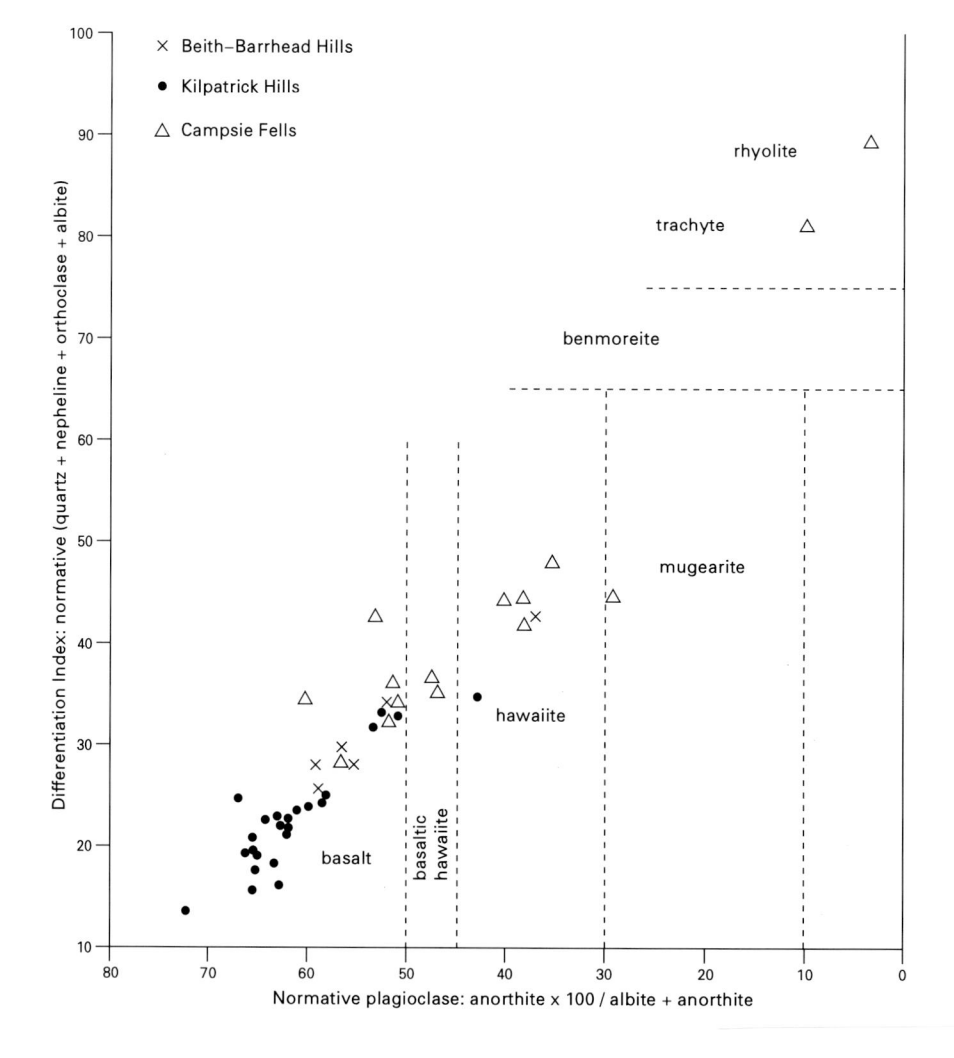

due to low-pressure fractionation in a long-lasting reservoir situated beneath it. The volcanicity is thought to be due to a rapidly propagating thermal cycle giving rise to magma pulses which, in turn, initiated regional tensional stress regimes. In this regime, normal faulting gave rise to differential relative subsidence of fault-bounded blocks, such as the Kilpatrick and Campsie blocks, within the Midland Valley graben. Smedley (1986) suggested that passive rather than active rifting took place. This suggestion is based on the absence of a hot-spot trail that would be expected if a deep mantle plume under a rapidly moving Midland Valley terrain was responsible for the magmatism.

Kirkwood Formation

The Kirkwood Formation consists of sandstone, siltstone, mudstone and some conglomerate composed of weathered detritus derived from the upstanding volcanic terrain formed by the Clyde Plateau Volcanic Formation. The strata are generally red, brown or purple in colour. In the Glasgow district, the formation mainly occurs at the surface south of the River Clyde, in the ground south of Johnstone and Paisley. Its thickness is variable and largely uncertain, but it may reach 30 m in this area.

North of the river it crops out around Douglas Muir [525 745]. To the west of Douglas Muir [508 748], the formation, which is about 15 m thick, overlies the Clyde Plateau Volcanic Formation and is in turn overlain by the Douglas Muir Quartz-Conglomerate Member. To the east of Douglas Muir [532 745], the relationship with the overlying conglomerate is the same but here, the Kirkwood Formation is underlain by fine-grained sandstone interbedded with seatearth of the Lawmuir Formation. Farther east and to the north on Craigmaddie Muir [565 780], the formation is absent and conglomerate lies directly on lavas. To the south, volcaniclastic rocks of the formation have been found in a few boreholes such as Lawmuir [5183 7310], where they are 15 m thick.

Generally, the Kirkwood Formation immediately overlies the lavas of the Clyde Plateau Volcanic Formation, from which it is clearly derived, and filled hollows in the top surface of the eroded and subsiding lava terrain. However, upstanding lava terrain occurred at intervals to supply volcaniclastic detritus, at least locally, after the onset of fluvial and fluviodeltaic sedimentation which laid down the Lawmuir Formation.

Lawmuir Formation

In most of the district, the Lawmuir Formation overlies the Kirkwood Formation. However, north of the River Clyde, the lower strata of the formation are interbedded or contemporaneous with the Kirkwood Formation in areas where volcaniclastic detritus was not being supplied to the basin.

The Lawmuir Formation belongs mainly to the Brigantian Stage of the Dinantian but some of the lowest beds may be Asbian in age. The top of the formation is now drawn at the base of the Hurlet Limestone, not the Hurlet Coal. The formation was previously known as the Upper Sedimentary Group of the Calciferous Sandstone Series or Measures. It is present over most of the central and southern parts of the district, but is little known except at or near its outcrops (Figure 9). The larger of these extends in a WSW–ENE belt across the centre of the district from Bishopton through Clydebank and Milngavie; the other occupies most of the Paisley area. At its type locality, the Lawmuir Borehole [5183 7310], the Lawmuir Formation is fully 250 m thick. This appears to be a fairly typical thickness, but at Paisley the formation reaches 330 m before thinning rapidly to 130 m at Johnstone and only 12 m at Howwood, just beyond the western margin of the district.

STRATIGRAPHY

The basal member of the Lawmuir Formation is the Douglas Muir Quartz-Conglomerate which is 20 m thick at the type locality [52 74]. It has been traced southwards as far as the Lawmuir Borehole (but is absent south of the River Clyde) and eastwards as far as Strathblane–Balmore Tunnel No. 8 Borehole [5713 7733]. This member consists of interleaved conglomerate and pebbly sandstone with a few lenses of mudstone. Trough cross-bedding is common in sets 0.2–1.0 m thick and many of the units are upward fining. The pebbles, which are almost exclusively of quartz and well rounded, are mostly 2 cm but cobbles up to 10 cm in diameter are also present. Palaeocurrent measurements on the cross-bedding show a wide spread with a vector mean to the SSW. This is compatible with derivation from the north as suggested by Tait (1973) based on a study of the heavy minerals in the overlying Craigmaddie Muir Sandstone. The conglomerate has been extensively quarried in recent years, on Douglas Muir itself and at Muirhouse [57 78].

The Douglas Muir Quartz-Conglomerate is succeeded north of the Clyde by the Craigmaddie Muir Sandstone which consists of white, cross-bedded sandstones, 170–180 m thick, with pebbly bands in places. These sandstones are prominently exposed from the type area at Craigmaddie Muir [590 765] north-eastwards to Lennox Castle [605 783] and to a lesser extent elsewhere. They are locally interbedded with mudstones and rare intercalations of volcanic detritus. North of the Clyde, the rest of the Lawmuir Formation is more varied in lithology but consists largely of sandstones similar to the Craigmaddie Muir Sandstone, with poorly bedded mudstones of the type sometimes called marls or fireclays (see below) and coals that are generally thin and poor in quality. The nonmarine Baldernock Limestone occurs near the top. Three marine bands, all with thin coals below them, are developed within bedded mudstones; the highest and middle also contain limestone beds. The Balmore Marine Band, the lowest of the three, has a relatively sparse fauna consisting mainly of brachiopods. It is known only from the Strathblane–Balmore Tunnel, the Lawmuir Borehole and two shallow boreholes in Milngavie. The other two marine bands, the Craigenglen Beds below and the Balgrochan Beds above, are much thicker and richer. The faunas of both include brachiopods and molluscs but large productoids and corals

28 FOUR LOWER CARBONIFEROUS (DINANTIAN)

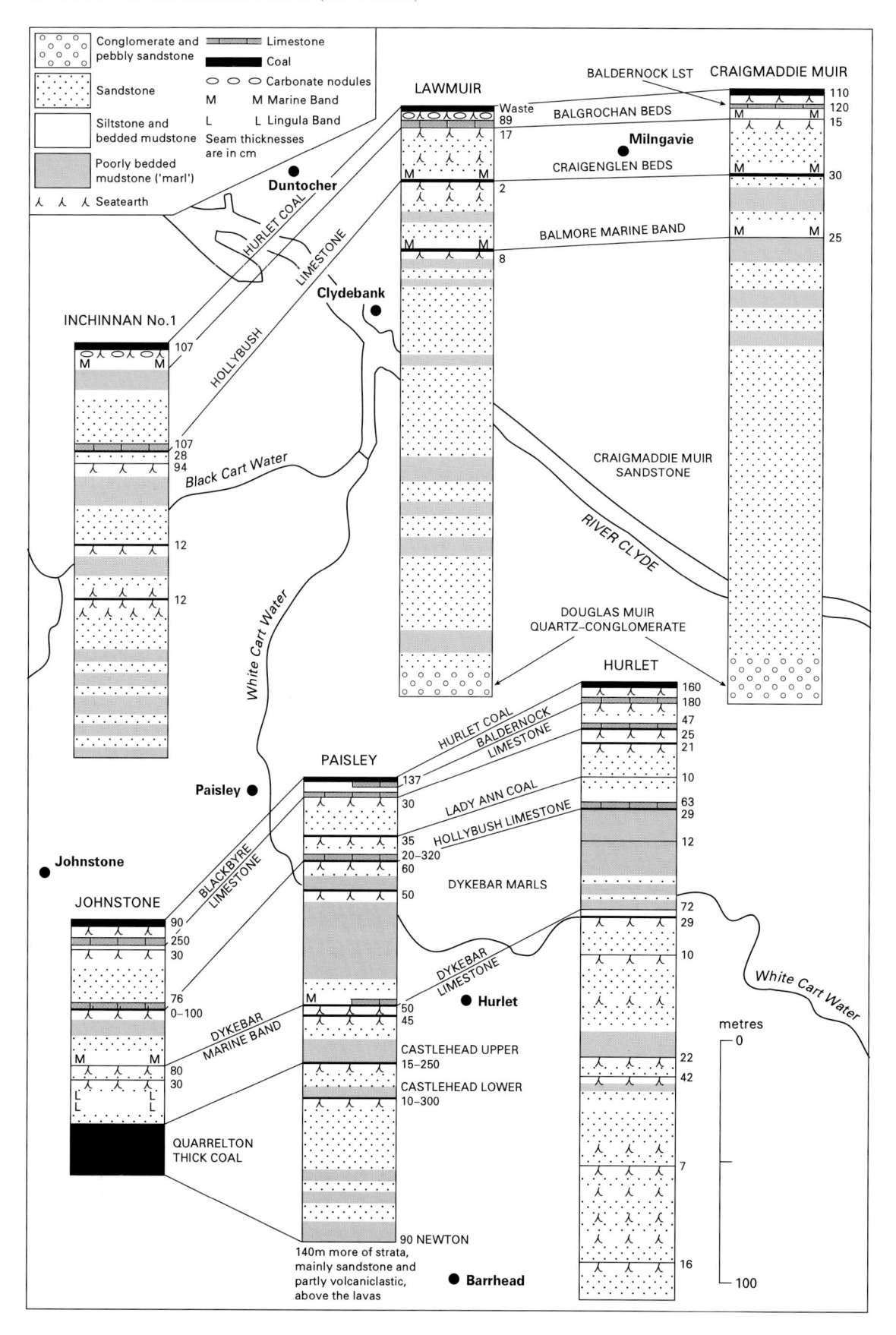

Figure 9 Comparative vertical sections of the Lawmuir Formation superimposed on a topographical map to show approximate sites.

are confined to the Craigenglen Beds. This feature strengthens the correlation of the latter with the Hollybush Limestone of the Paisley area, a horizon also recognised at Duntocher. The Craigenglen Beds are well exposed in a small stream near Baldernock Mill [5748 7484], at Glenwynd [6093 7588–6100 7581] and in several streamlets south of Lennox Castle including Baldow Glen [6160 7770]. The Balgrochan Beds are also exposed in Glenwynd [6085 7598] and in the streamlets south of Lennox Castle including Baldow Glen [6153 7753], as well as in Barraston Burn [6013 7080].

The nonmarine Baldernock Limestone is up to 3.5 m thick and lies just above the Balgrochan Beds. It was also known as the White Limestone on account of its pale colour. Because of its impure dolomitic nature, it has not been exploited to any extent but it was mined using stoop-and-room methods at the type locality [591 757], the Linn of Baldernock (Plate 5), where it has been baked by a dolerite sill. The old workings there are still visible. The fauna consists of *Carbonita* sp. and fish debris. Roots also occur, showing that the limestone formed part of the seatbed of the overlying Hurlet Coal. Indeed in the Lawmuir Borehole the root penetration and leaching were so intense that the limestone is reduced to carbonate nodules in seatclay.

North of the River Clyde, the Hurlet Coal is the thickest (up to 1.8 m) in the Lawmuir Formation and the only seam known to have been mined. Workings were extensive from Duntocher to the area south of Lennox Castle known as the South Brae of Campsie. However, there are no workings between Milngavie and Balmore, where the coal has been destroyed by the Milngavie dolerite sills, as can be seen at the Linn of Baldernock.

South of the River Clyde (Figure 9), the Douglas Muir Quartz-Conglomerate is absent and there is no single development of sandstone comparable to the Craigmaddie Muir Sandstone. However, in the Paisley area, where the formation is 240 m thick, the lower part of the sequence (up to the Dykebar Limestone) contains many sandstones similar to those found at Craigmaddie Muir. There is also a greater proportion of poorly bedded siltstone and mudstone, some intercalated volcaniclastic material, especially south of Johnstone, and a variable number of mostly thin coals. A similar sequence, 170 m thick, was found in the Hurlet Borehole [5111 6125]. The mudstones mostly contain bands and clusters of sphaerosiderite, making them locally rich enough in iron to be formerly worth prospecting as ironstones. However, the only known attempt to mine ironstone, at Glenburn [4768 6138], south of Paisley, was short-lived. Three of the mudstones were worked for fireclay at the Caledonia Fireclay Pit [4751 6495]; two lay immediately above and below the Castlehead Lower Coal and the third 27 m above this seam. Five of the coals in places attain workable thicknesses. The lowest is the Newton Coal, almost 100 m below the Dykebar Limestone, which is up to 0.9 m thick and was mined to a small extent at the type locality [4570 6276]. The Castlehead Lower and Upper coals are poor in quality and very variable in thickness, up to 3.0 m for the former and 2.5 m for the latter. Both have been mined to some extent, for instance from Castlehead Pit [4746 6346]. The Lower and Upper Dyke-

Plate 5 Stoop and room workings in the Baldernock Limestone (Lawmuir Formation), Linn of Baldernock (D 3369).

bar coals are up to 0.6 m and 0.75 m thick respectively; recent boreholes suggest the lower one was worked in the vicinity of Ferguslie Fireclay Works [461 634]. East of Paisley all these coals deteriorate; in the Hurlet Borehole none was more than 0.42 m thick. West of Paisley the Newton and Castlehead coals unite to form the Quarrelton Thick Coal (see Carruthers in Hinxman et al., 1920, pp.19–28, for a detailed description). Locally this seam exceeds 20 m in thickness and is said to reach 30 m as a result of a contemporary bog-slide, but it is very variable and subject to 'wash-outs'. The seam has been much exploited since the 18th century or earlier. South of Johnstone, volcaniclastic sedimentary rocks alternate with coal seams which are probably components of the Quarrelton Thick Coal. Local names were used at the Wallace Pit, Elderslie [4391 6283] for such components of the thick coal (in descending order): Nine-Foot, Smiddyhead, Ironhead, Merryhead, Minehead, Craighead.

The upper part of the Lawmuir Formation, starting with the Dykebar Limestone, is 80–90 m thick. Sandstones are less dominant and poorly bedded mudstones ('marls' or 'fireclays') form a considerable part of the sequence; they are usually fawn or pale grey with red, purple, green or yellow mottling, are locally rich in sphaerosiderite and are probably partly volcaniclastic in origin. The Dykebar Marls, which lie between the Dykebar and Hollybush limestones, are up to 40 m thick. The Ferguslie Fireclay was mined at the Ferguslie [4604 6346] and Caledonia fireclay pits. At the latter three other fireclays lying between the Sandholes and Hollybush coals were also exploited. There are several coals, the main ones beings the Sandholes (0–0.7 m thick), Hollybush (0–1.15 m), Lady Ann (0–0.46 m) and Hurlet (generally 0.9–1.75 m). The Lady Ann Coal was worked to a small extent from a pit [5008 6002] north of Barrhead. The Hurlet Coal was extensively extracted many years ago, especially in the Hurlet and Nitshill areas (except where it was destroyed by a dolerite sill) and locally elsewhere. It is dirty and pyritous but has good coking properties.

The sequence includes three marine limestones and their associated bedded, usually calcareous mudstones. The lowest limestone, the Dykebar, is impersistent, argillaceous and up to 1.4 m thick. The associated marine mudstones are much more persistent and therefore the horizon of the Dykebar Marine Band can be traced over the southern part of the Paisley area, although it was absent in Inchinnan No. 1 Borehole [4521 6888]. The fauna of the Dykebar Marine Band is relatively sparse and consists mainly of bivalves and brachiopods. Until recently it was thought to be the lowest marine band in the Lawmuir Formation but it is now known that at least two minor marine incursions took place a little earlier. Thin mudstones with *Lingula* sp. are present in the Johnstone area about 12 m and 16 m respectively below the Dykebar Marine Band.

The Hollybush Limestone and its associated mudstones have a richer and more varied fauna than the Dykebar Marine Band, including abundant brachiopods and corals (mainly *Siphonodendron* sp.) but relatively few molluscs. The limestone occurs throughout the Paisley area and is up to 5 m thick. It was formerly quarried at Hollybush [494 612], Elderslie [447 631], Brediland [460 629], Nethercraigs [465 610; 468 610] (where it is still visible), Brownside [488 605] and Garnieland [480 696]. Another exposure is in the railway cutting [488 633] at Saucel Hill.

The Blackbyre Limestone and its associated mudstones also have a persistent and diverse fauna with rich assemblages of brachiopods and corals including *Diphyphyllum* sp. East of Paisley (Figure 9) the marine beds are separated from the Hurlet Coal by 12–15 m of strata, including the nonmarine Baldernock (or White) Limestone, which is up to 2.85 m thick with a fauna of ostracods. West of Paisley the intervening beds, including the Baldernock Limestone, die out and some or all of the marine strata, including the Blackbyre Limestone, form part of the seatbed of the Hurlet Coal. The resultant leaching locally reduces the limestone to nodules in a greenish grey clayey mudstone (Forsyth and Wilson, 1965, pp.68–69). Carruthers (in Hinxman et al., 1920) regarded such a development as characteristic of the Baldernock Limestone and mistakenly believed that the latter was marine west of Paisley. The leaching, and the fact that many of the mudstones are sufficiently calcareous to be verging on argillaceous limestone, make the descriptions of the marine beds in many of the older boreholes difficult to assess. In consequence it is difficult to determine their thickness and that of the Blackbyre Limestone itself, which is said to reach 7 m at Nethercraigs, where it is still partially exposed. The total thickness of the marine strata is generally between 5 m and 10 m, but in Crookston No. 1 Borehole [5198 6195] it is only 4 m and the limestone is only 0.35 m thick. This may indicate eastward deterioration of this marine phase.

Another marine episode began just before the end of the deposition of the Lawmuir Formation and resulted in the deposition of the Alum Shale, up to 1 m thick, between the Hurlet Coal and the Hurlet Limestone.

PALAEOGEOGRAPHY

The Lawmuir Formation was laid down on an irregular and variably subsiding land surface of volcanic rocks. The hollows in this surface in the south had been largely filled by the volcaniclastic sedimentary rocks of the Kirkwood Formation but in the north only to a small extent. Volcaniclastic material continued to accumulate in the south, alternating with quartzose detritus mainly of sand grade and fluvial origin. In the north, quartzose detritus dominated for a long period and initially included many quartz pebbles, probably of Highland origin. For a long time neither colonisation by plants nor invasion by the sea occurred in the north. However, in the south, coal swamp conditions became established early and were repeated several times. Near Johnstone during this period, a small, possibly fault-controlled basin was developed. Sediment was largely excluded from the basin by the continuous presence of raised mires leading to the formation of the exceptionally thick seam, the Quarrelton Thick Coal. Shortly after that two quasimarine incursions took place in the south-west. The first major marine incursion followed and covered most of the southern part of the district, but the fauna was restricted and sufficient mud was available so that only locally did even impure limestone form. Mud and silt, at least partly of volcanic origin, continued to be deposited at intervals especially in the south, probably as overbank deposits between the now sand-filled channels. Colonisation by vegetation occurred at intervals but only the final episode was prolonged. Two other marine invasions took place, bringing in rich faunas and clear waters particularly in the south. The second was followed by an episode of nonmarine limestone formation except in the south-west. The base of the limestone deposited during the next marine incursion is taken as the end of the Lawmuir Formation.

CLACKMANNAN GROUP (DINANTIAN PART)

Lower Limestone Formation

The Lower Limestone Formation forms the highest part of the Brigantian sequence and belongs to the VF Miospore Zone. It is present over most of the southern part of the district. The formation thickens markedly eastwards from 60 m near the western margin of the district to a reported maximum of 179 m in an underground borehole [5694 6264] at Titwood No. 2 Pit. It also thickens southwards from 110 m in Balmore No. 2 Borehole [6013 7468] to 165 m in Milngavie No. 5 Borehole [5735 7285]. A detailed description of the stratigraphy is given in Forsyth (1993).

LITHOLOGY

The Lower Limestone Formation (Figure 10) consists largely of bedded, dark grey mudstones. They form almost all of the thinner sequences in the west and north (Forsyth and Wilson, 1965), and also most of the formation in the Titwood Borehole. Siltstone and sandstone contribute up to 25 per cent of the sequence, as in Milngavie No. 5 Borehole. These coarser beds occur in four parts of the succession. The lowest, 14–20 m above the Hurlet Limestone, has been recorded only in the Titwood Borehole. The next, 8–21 m below the Blackhall Limestone, is present only in the north-east. The thickest (up to 31 m) lies between the Milngavie Marine Band and the Main Hosie Limestone and includes the Hosie Sandstone, which is 20 m thick in Milngavie No. 5 Borehole but is absent in the west, where it is represented by siltstone. Like the Hosie Sandstone, this siltstone locally contains some marine fossils, mostly brachiopods, an unusual feature for a Scottish Carboniferous sandstone. All these sandstones grade upwards from mudstone, through siltstone. The highest sandstone, known as the Lillie's Sandstone, has a rather different and more variable development. It occurs intermittently, associated with the Lillie's Shale Coal, between the Mid and Second Hosie limestones, and exceeds 13 m in thickness in Milngavie No. 2 Borehole [5392 7357]. Coals or seatearths locally occur both above and below this sandstone, for instance in the Titwood Borehole it is 7 m thick with 0.5 m of coal below it and 0.66 m above. The Lillie's Shale Coal is the only one known to have been mined in the Lower Limestone Formation in the district. In the Paisley area, where it was exploited for both coal and oil shale, it typically consisted of 0.3 m of coal on 0.35 m of oil shale; elsewhere it is much more variable and generally thinner.

Limestones The limestones which give the Lower Limestone Formation its name are mostly marine, crinoidal and up to eight in number. They are stratigraphically important but thin. The only nonmarine limestones are ostracod-rich irony limestones below the Blackhall Limestone. Most of the limestones are thin and only locally exceed 1 m in thickness. The exception is the Hurlet Limestone which shows a marked increase to 11 m at Nethercraigs (or Limecraigs) [466 601], south of Paisley, where it was quarried. This thickness, the maximum in the district, is paralleled at Howwood in the Greenock district (Forsyth, 1978a) and at Lennoxtown in the Airdrie district (Forsyth et al., 1996). These thick developments of limestone are at the present limits of outcrop of the formation, in areas where the overall sequence is unusually thin and probably near the margin of the depositional basin. The Hurlet Limestone was formerly quarried and mined, together with the underlying alum shale and coal, around Hurlet itself, Duntocher, Baljaffray [533 737] and Newlands [607 766]. None of the other limestones is known to have been worked in the district. The Top Hosie Limestone, which is very argillaceous and less abundantly crinoidal than the others, was formerly thought to have been quarried as a cementstone in the Paisley district but this is doubtful.

Mudstones The mudstones above the limestones are also partly marine but the richness and variety of the faunas is notably variable. The mudstone above the Hurlet Limestone is sparsely fossiliferous. The best is in the Neilson Shell Bed above the Blackhall Limestone, whose fauna is distinctive and unique in the Glasgow district (Wilson, 1966, p.111); characteristically it contains *Crurithyris urii, Tornquistia youngi, Glabrocingulum atomarium, Straparollus carbonarius, Euchondria neilsoni, Pernopecten fragilis* and *Posidonia corrugata gigantea* in a rich fauna of brachiopods and molluscs including goniatites, plus a few solitary corals. The mudstones associated with the Hosie Limestones also contain rich and varied faunas, mainly of brachiopods, gastropods and bivalves. However, most of the remaining mudstones, particularly in the lower part of the formation, are either barren or yield only ostracods (locally in great abundance, for instance under the Blackhall Limestone), fish debris or *Curvirimula*.

Clayband ironstones are abundant in the mudstones, especially below the Blackhall Limestone and between it and the Milngavie Marine Band (Figure 10). The lower ones are known as the Lower and Upper Househill clayband ironstones, about 12–16 m and up to 6 m below the Blackhall Limestone. Individual ironstones reach 0.35 m in thickness. They were formerly mined at several pits in south-west Glasgow. North of the Clyde, the

FOUR LOWER CARBONIFEROUS (DINANTIAN)

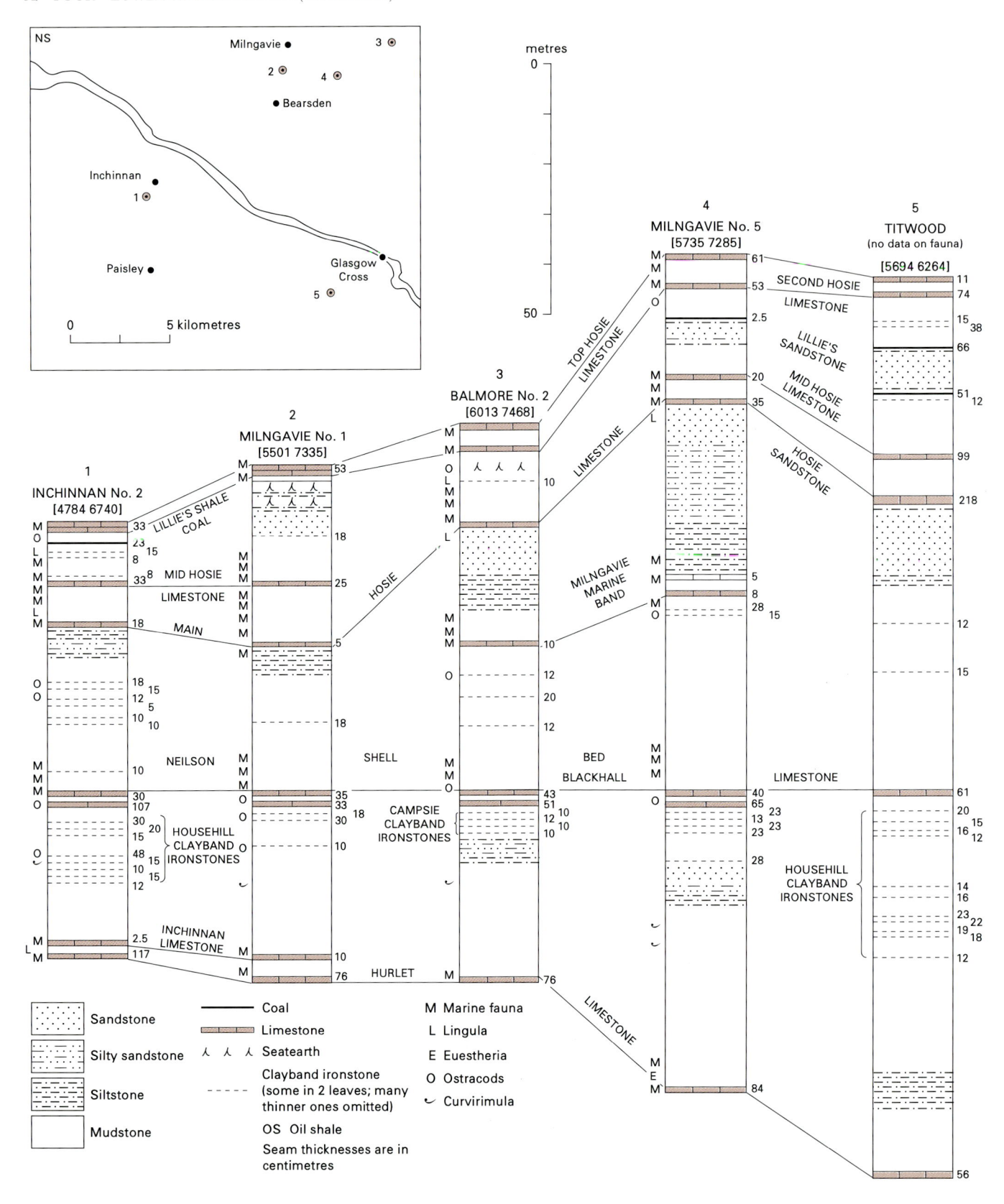

Figure 10 Comparative borehole sections to show variation in thickness in the Lower Limestone Formation.

Upper Househill Clayband Ironstones were known as the Campsie Clayband Ironstones and were worked at Blairskaith [596 752].

PALAEOGEOGRAPHY

The marine transgression that produced the Hurlet Limestone marked a significant change in the sequence. Below it are the sandstones, coals and poorly bedded, generally pale grey mudstones of the Lawmuir Formation, with rare marine bands composed of very shelly, calcareous mudstones, with abundant brachiopods, and locally coraliferous limestones. Above it are the bedded, characteristically dark grey mudstones, with clay ironstone bands, siltstones, silty sandstones and thin crinoidal limestones of the Lower Limestone Formation, with either sparse nonmarine faunas or marine faunas dominated by molluscs. This was the time when the district was a quiet backwater, subsiding slowly but steadily as fine sediment was gradually brought in, probably mainly from the east. In that direction the formation continues to thicken and sandy lobes extended westwards into the district from a large delta that is thought to have occupied much of the central part of the Midland Valley. This delta probably limited access by the sea to much of the district and may have restricted water circulation sufficiently to produce anaerobic conditions in which few benthonic forms other than ostracods could survive, except when increased subsidence led to inundation of the whole delta. The margins of the depositional basin to the north, west and south probably lay close to the district for much of the time, but the neighbouring land masses were so peneplaned that they contributed little detritus to the slowly subsiding basin.

FIVE

Upper Carboniferous (Silesian)

The Upper Carboniferous strata are assigned to two groups. The lower is the upper part of the Clackmannan Group which comprises the Limestone Coal Formation, the Upper Limestone Formation and the Passage Formation. The upper is the informal group, the Coal Measures which consists of the Lower, Middle and Upper Coal Measures. The boundary between the two lithostratigraphical divisions is close to the Namurian and Westphalian boundary. More detailed descriptions of the stratigraphy are given in Forsyth (1993).

CLACKMANNAN GROUP (SILESIAN PART)

The Silesian part of the Clackmannan Group is characterised by the continuation of the deposition of cyclothemic sedimentary rocks which started in the upper Dinantian. All three formations consist of sandstone-dominated cycles with siltstones, mudstones, limestones and coals occurring in different proportions in each formation. The top of the Limestone Coal Formation is taken at the base of the Index Limestone, the top of the Upper Limestone Formation at the top of the Castlecary Limestone, and the top of the Passage Formation at the base of the Lowstone Marine Band.

Limestone Coal Formation

The Limestone Coal Formation is the lowest part of the Namurian sequence and belongs entirely to the Pendleian (E_1) Stage. It occurs in most of the south-east part of the district and also on the north-west side of the Paisley Ruck from Johnstone to Renfrew. The formation can be divided into two parts at the top of the Black Metals, a development of mudstones about 20 m thick.

LITHOLOGY

The lower part of the formation (Forsyth, 1978c) consists of bedded dark grey mudstones, similar to those in the Lower Limestone Formation, with siltstones, sandstones, numerous clayband ironstones and up to five blackband ironstones or coals near the top (Figure 11a). The upper part (Forsyth, 1980) is a coal-cyclic sequence (Figure 11b). The cycles are up to 25 in number and average 6–7 m in thickness. They have been traced from Glasgow to Stirling (Forsyth and Read, 1962). The cycles usually consist in upward sequence of coal, mudstone, siltstone, sandstone (generally the thickest item), siltstone, seatearth and coal (of the next cycle). The cycle that starts with the Glasgow Shale Coal is usually the thickest (10–15 m). Many of the coals pass laterally wholly or partly into blackband ironstones. Despite its name, the Limestone Coal Formation contains few limestones, the only ones being local bands of argillaceous shelly limestone in the Johnstone Shell Bed and the impersistent nonmarine Berryhills Limestone. The lower part of the formation thickens notably eastwards from 125 m at Linwood to 177 m in the Colston Road Borehole [5945 6913]. The upper part also thickens eastwards from 145 m to 170 m, and the whole formation from 270 m to 340 m.

Sandstones The sandstones are mostly fine grained, with sideritic micaceous ripple-laminae and silty laminae towards the top and base. Conspicuous bioturbation is locally displayed, notably by *Teichichnus* above the Kilsyth Coking Coal. The sandstone above the Knightswood Gas Coal thickens from 4 m to 10–15 m in the W–E belt through north-central Glasgow from Partick through Hillhead and Woodside to Robroyston (Sheet 31W). This thickening does not affect the underlying coal but has a marked effect on the strata above, which are unusually thin, silty and rooty. The Jubilee and Glasgow Shale coals are thin or absent, but the *Lingula* band above the former is not affected. Exceptionally, in the Willowbank Crescent Borehole [5764 6655] the sandstone was 25 m thick and the Jubilee horizon was entirely absent. The underlying thick body of incompactible sand clearly disrupted the normal cyclic pattern but, except at its thickest, it subsided sufficiently to allow quasimarine conditions to extend. Medium- and coarse-grained sandstones with markedly erosive bases are uncommon. The Cowlairs Sandstone at the top of the formation is best developed north of the River Clyde, locally exceeds 20 m and was formerly quarried [541 699–544 697]. It cuts out the Twechar Dirty Coal and locally the Twechar Upper Coal. The Nitshill Sandstone occupies a sinuous belt 1.5–2 km wide, which extends from Barrhead in a WSW–ENE direction through Nitshill, where it was quarried, to central Glasgow and into the Airdrie district. At its maximum of about 20 m this sandstone replaces the Ashfield Rider and Fourteen-Inch Under coals, with the latter absent over a larger area. This suggests that the sandstone cut out the lower coal over the whole extent of the channel but inhibited formation of the higher coal only in the central part. This inference is strengthened by the presence at the margin in central Glasgow of a thin but coarse-grained sandstone between the two coals. Other unusually thick, coarse-grained sandstones include one in the Balmore area that cuts out the Berryhills horizon; others above the Possil Main and Possil Wee coals which occur locally in north-central Glasgow and several between the Giffnock Main and Possil Rider coals in the Darnley Basin.

Coals Below the Black Metals only the Kilsyth Coking Coal has been mined, to a limited extent. Above the

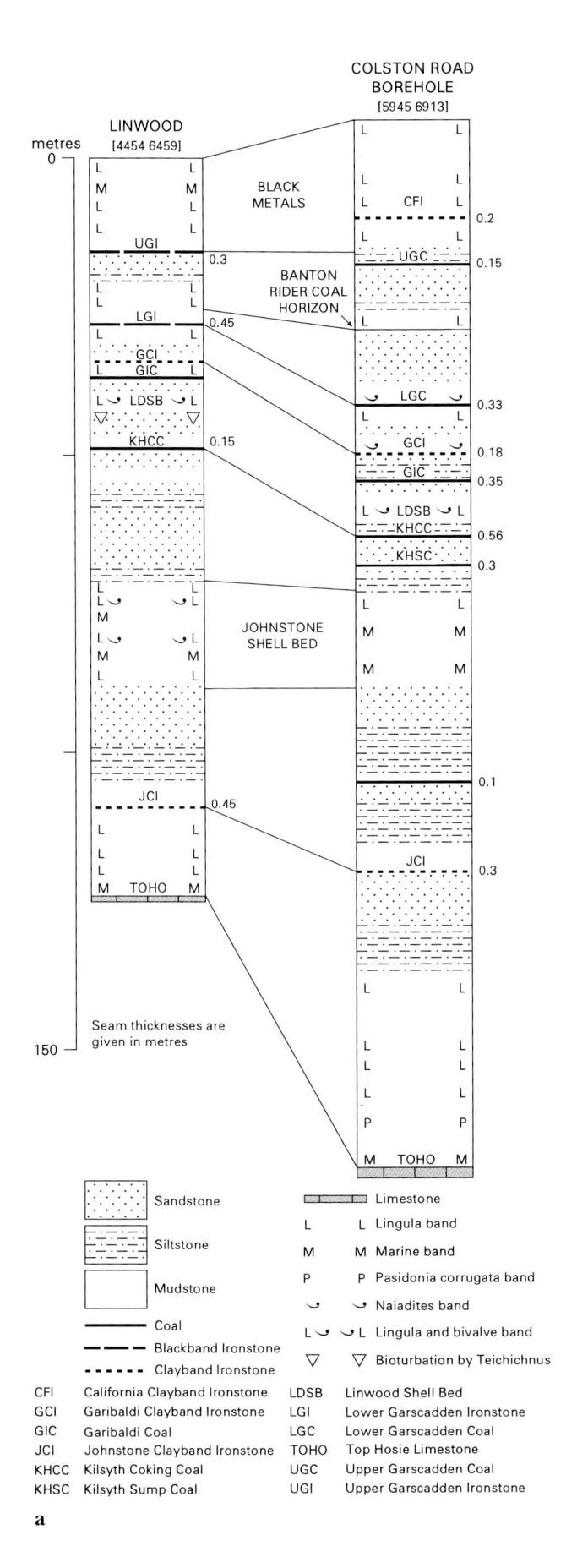

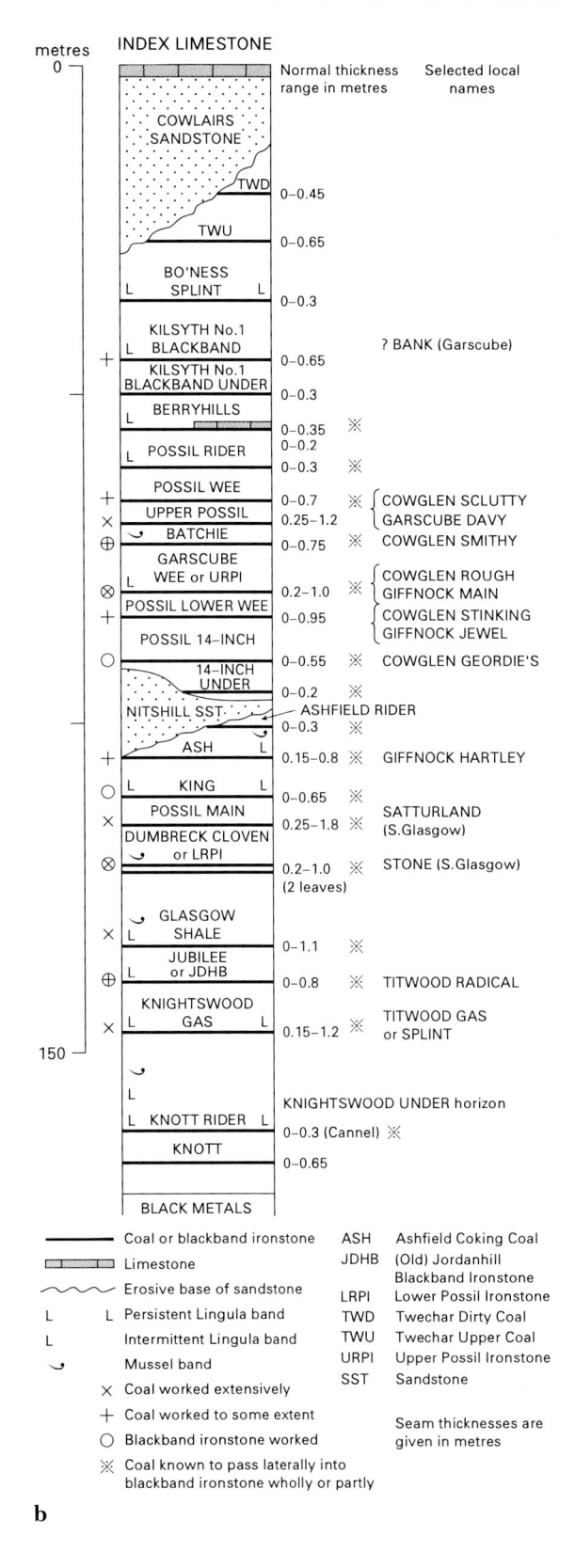

Figure 11 Vertical sections of the Limestone Coal Formation.

a. Comparative sections of the lower part showing eastward thickening.

b. Generalised section of the upper part.

Black Metals at least 12 coals have been mined, some extensively (Figure 11; Plate 6). The Knightswood Gas, Glasgow Shale, Possil Main and Upper Possil coals locally exceed 1 m and the Possil Main is reputed to reach 1.8 m at Knightswood. Otherwise the coals are generally less than 1 m thick.

Clayband ironstones   Three clayband ironstones (Johnstone, Garibaldi and California) have been mined. The first-named, in two leaves totalling 0.45–0.6 m, was worked mainly south of the Clyde, from Johnstone to south-west Glasgow, and also in the north-west part of the city. The Garibaldi seam, also in two leaves 0.3–0.4 m in total, was extensively extracted from north-west and north Glasgow, and to a small extent in south-west Glasgow. The California Ironstone, also in two leaves 0.3–0.5 m thick, was worked to a limited extent in north-west and north Glasgow.

Blackband ironstones   It is for blackband ironstone, however, that the Limestone Coal Formation of the district is particularly notable. Material of this type (interleaved clay ironstone and coal, usually canneloid) locally replaces wholly or partly 18 coals. Nine of these ironstones are known to have been mined, five of them extensively, namely the Lower and Upper Garscadden, Old Jordanhill Black-

Plate 6a. Shallow workings in the Giffnock Main Coal (Limestone Coal Formation) showing stoops and rooms, Thornliebank (C 3592).

Plate 6b. Stoop-and-room mining in the Bishopbriggs Sandstone (Upper Limestone Formation) Huntershill, Bishopbriggs, north Glasgow (C 2418).

band, Lower and Upper Possil seams. (The New Jordanhill Blackband Ironstone is a local phase of the Garibaldi Coal.) This is the greatest concentration of blackband ironstone development known in the Limestone Coal Formation anywhere in the Midland Valley, with its maximum in north-west Glasgow. Some, such as the Lower and Upper Garscadden ironstones, the Old Jordanhill Blackband Ironstone and the Batchie Ironstone, replace the whole coal seam. In the Upper Possil Ironstone only the top part becomes ironstone, and in the Lower Possil Ironstone only the upper leaf of the Dumbreck Cloven Coal is replaced.

Marine bands   The Limestone Coal Formation contains two major marine bands, namely the Johnstone Shell Bed and the Black Metals Marine Band. Both occur throughout the district. The Johnstone Shell Bed is by far the richer (Wilson, 1967). It occurs from Johnstone to Glasgow in a single bed of mudstone over 20 m thick, in which two bands containing rich faunas of brachiopods, bivalves and gastropods (but few cephalopods or corals) alternate with bands containing *Lingula* either alone or with *Paracarbonicola* and *Naiadites*. North of Glasgow the mudstone splits, with a marine fauna in the lower leaf and *Lingula* in the upper one. The fauna of the Black Metals Marine Band is not as rich. It includes brachiopods, gastropods and bivalves, a few cephalopods but no corals.

The mudstones immediately above the Top Hosie Limestone contain a restricted marine fauna including abundant *Posidonia corrugata*; the overlying mudstones contain several *Lingula* bands, in some of which *Naiadites* also occurs (Forsyth and Wilson, 1965). The Linwood Shell Bed contains the *Lingula–Paracarbonicola–Naiadites* fauna noted above, but rarely, with a few other bivalves. *Lingula* bands occur above the Garibaldi and Banton Rider coals and below the Lower Garscadden Ironstone. (Only *Naiadites* occurs immediately above the last of these seams.) In addition to the Black Metals Marine Band, the Black Metals contain up to four *Lingula* bands and *Naiadites* also occurs.

Lingula bands are numerous above the Black Metals (Forsyth, 1980) especially in northern Glasgow, which has the maximum number (12) for this part of the succession in the Midland Valley. Some of the bands disappear or thin out towards the north-west, west or south-west. In most, only *Lingula* is present but fish debris, ostracods and *Curvirimula* fragments also occur locally. In some bands, particularly those in blackband ironstones such as the King, the *Lingula* itself is also fragmentary, suggesting much posthumous winnowing and transport. Nonmarine bivalves are uncommon above the Black Metals but *Naiadites* occurs abundantly in the Lower Possil and Batchie ironstones (to the exclusion of *Lingula*), and *Naiadites* and *Curvirimula* have been found locally in the thick mudstone at the Knightswood Under horizon, above the Glasgow Shale and Ashfield Coking coals, and rarely elsewhere.

PALAEOGEOGRAPHY

When the lower part of the Limestone Coal Formation was laid down, the quiet backwater conditions of the Lower Limestone Formation continued to prevail, especially in the west. Periodically, stronger currents brought in silt and sand down the frontal slope of the delta that lay to the east and largely inhibited fully marine conditions from developing. At all times it was too muddy for corals. Quasimarine conditions occurred frequently in which *Lingula* flourished, sometimes along with a restricted range of bivalves. The waters were periodically saturated with iron leading to the development of bands and nodules of clay ironstone in or on the muddy substrate. During deposition of the upper part of the sequence, the land surface periodically emerged and vegetation colonised the alluvial plain leading to the local development of thin coals. Elsewhere, iron precipitation and the accumulation in shallow boggy stagnant waters of mud very rich in decaying vegetation resulted in the formation of blackband ironstones. Finally mud deposition again dominated, and the Black Metals were formed.

The upper part of the formation shows a change to paralic deltaic conditions as the eastern delta extended over the district. Quasimarine conditions however frequently reached all or part of the Glasgow district and boggy areas of iron-rich stagnant water formed at frequent intervals on top of the delta to receive finely comminuted vegetable matter when elsewhere coal-forming vegetation colonised the emergent alluvial plain. Occasionally river channels cut across the plain and laid down coarser-grained sandstones.

Upper Limestone Formation

The Upper Limestone Formation crops out in a sinuous belt from Barrhead in the south-west through central Glasgow northwards to Balmore and thence south-west to north-west Glasgow, and also in a few outliers about which little is known. It belongs partly to the Pendleian (E_1) Stage of the Namurian and partly (from the Orchard Limestone upwards) to the Arnsbergian (E_2) Stage. These two parts (Forsyth, 1982) show contrasting successions (Figure 12).

LITHOLOGY

The lower (120 m thick) starts with the marine Index Limestone and its overlying mudstones, the Bishopbriggs Sandstone and the Huntershill Cement Limestone. The rest of the lower part consists largely of coal-cyclic sequences, except for the marine Lyoncross Limestone and associated mudstones. The cyclic sequences are, however, extensively replaced by thick, medium- to coarse-grained sandstones. The upper part starts with two major cycles of Yoredale type that begin with the Orchard and Calmy limestones and their thick overlying mudstones. The lower cycle is 55 m thick, the upper 45–50 m. Above, there are some 35 m of alternating strata including two thin limestones, an 11 m sandstone and the Castlecary Limestone at the top. The highest beds occur only in the north, where they escaped later Namurian erosion, and the formation reaches its greatest thickness in the district of 285 m in Cadder No. 6 Borehole [5876 7318]. In the Darnley area, the preserved thickness is about 250 m.

Limestones   The limestones that give the formation its name form only about 3 per cent of the sequence but are

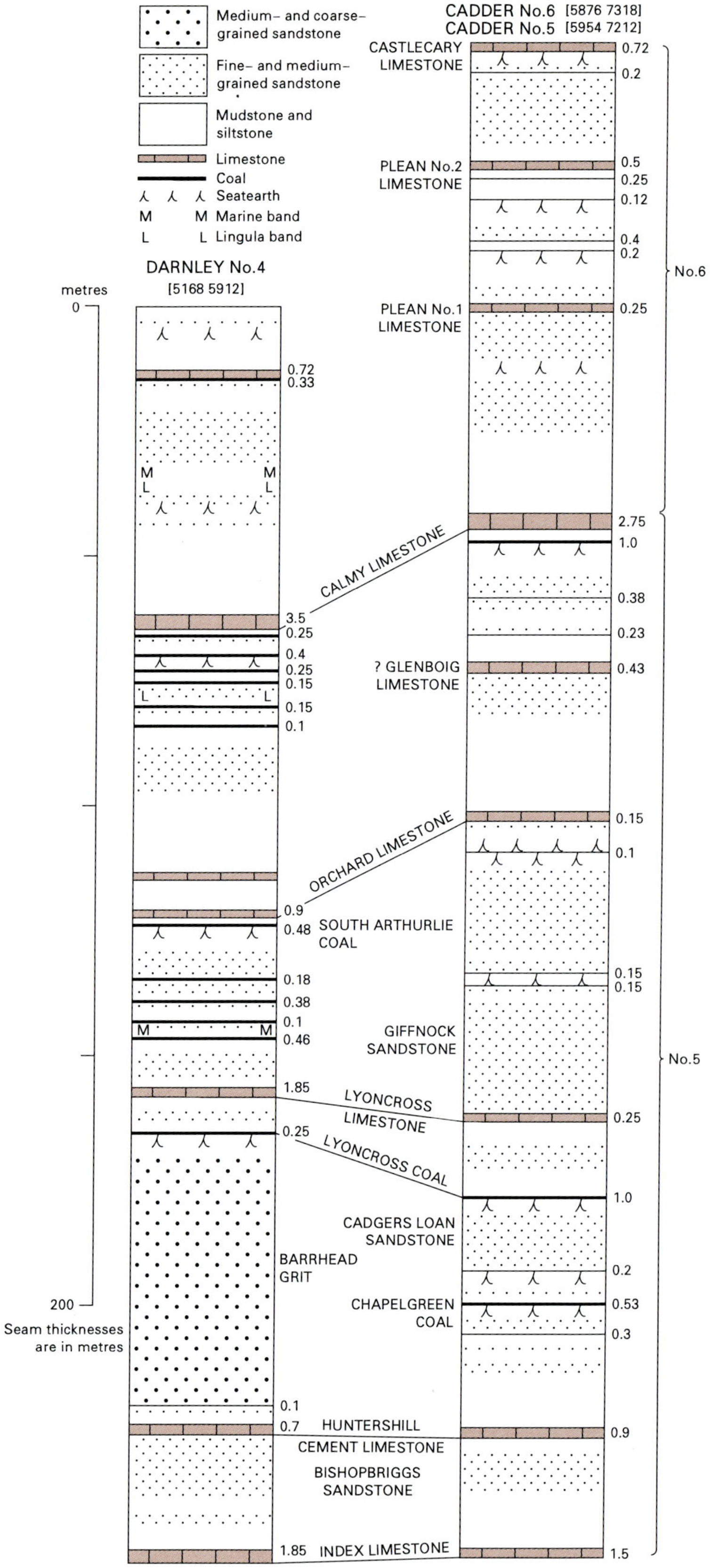

Figure 12 Comparative borehole sections in the Upper Limestone Formation.

important biostratigraphically; some were formerly important economically. They are all crinoidal and thickest in the Darnley Basin; here the Index Limestone is normally 1.2–1.5 m thick but locally exceeds 2 m. It can be seen in the railway cutting at Bishopbriggs [603 694]. The Huntershill Cement Limestone (up to 1.15 m thick, normally in two beds) and mudstones contain an impoverished fauna. Both these limestones are locally cut out by the Upper Dumbreck Sandstone. The Lyoncross Limestone is 0.15–1.8 m thick and in two beds in the Darnley area. It was worked in Waulkmill Glen [522 581] and can still be seen there and in a railway cutting [530 601].

The Orchard Limestone is 0.15–1.2 m thick. It was quarried and apparently mined at Giffnock. In the Darnley area the Upper Orchard Limestone, up to 1.4 m thick, occurs in the thick overlying mudstones, which contain the 'richest, best-preserved and most varied fauna in the Scottish Namurian' (Wilson, 1967, p.458), including corals, polyzoa and trilobites as well as many brachiopods and molluscs. The Calmy Limestone (1.5–4.2 m thick) occurs in two beds. The lower is typically crinoidal but the upper is of 'calmy' type being fine grained, homogeneous and off-white, with conchoidal fracture and less obviously fossiliferous. The Calmy Limestone was formerly quarried and mined at Darnley [5235 5885] and is still exposed in Waulkmill Glen to the south. The associated mudstones contain a brachiopod-dominated fauna that is almost as rich as that found above the Orchard Limestone.

The Plean limestones are poorly known. No. 1 Limestone is 0.15 to 0.3 m thick at Townhead [602 659] and 0.25 m thick in Cadder No. 6 Borehole. It may be represented by a marine mudstone in the Darnley area. The 'extra' limestone of eastern Glasgow (Forsyth, 1961, p.219) probably occurs as a dolomitic bed 0.9 m thick at the bottom of the Govan No. 5 Pit Underground Borehole [6039 5240], as a limestone 0.85 m thick in Hampden Park No. 11 Borehole [5908 6139] and as the 0.33 m 'cement' formerly exposed at Garngad [6135 6655]. No. 2 Limestone has possible representatives in Cadder No. 6, Darnley No. 4 and Govan No. 5 Pit boreholes. No. 3 Limestone is absent, probably removed by the erosion that also locally removed No. 2 Limestone. The limestone, regarded as the Castlecary in Cadder No. 6 Borehole, was 0.72 m thick but its nature is otherwise unknown.

Marine bands   The Upper Limestone Formation in the district also contains a few minor marine or *Lingula* bands, but not as many as are known in the Airdrie district (Forsyth et al., 1996). Lack of opportunities to look for these bands and the presence of erosive sandstones may partly explain their relative absence but probably the marine bands tend to die out westwards. The lowest is the limestone, 0.3–0.38 m thick, within the Bishopbriggs Sandstone, that is known only in and beside the railway cutting [6048 6965] at Bishopbriggs. The Glenboig Marine Band (Forsyth, 1978b) has not been proved to occur in the district. It is absent in the Darnley area but is probably present around Cadder where several non-examined boreholes recorded both mudstone and limestone (probably the Glenboig Limestone) in the appropriate part of the sequence. A *Lingula* band occurs 10–15 m below the Calmy Limestone in the Darnley area. Between Plean Nos 1 and 2 limestones, two bands with *Lingula* and marine bivalves occur at Townhead [6023 6590] and three *Lingula* bands occur in Govan No. 5 Pit Underground Borehole. Five metres of sandstone and siltstone immediately above Plean No. 1 Limestone at Townhead are conspicuously bioturbated. Records of nonmarine bivalves are very rare but there are sufficient examples of *Naiadites* or *Curvirimula* in northern Glasgow at or near the Chapelgreen Coal to suggest that they occur quite commonly at that horizon.

Coals Most of the coals in the Upper Limestone Formation are less than 0.5 m thick. The Chapelgreen Coal is up to 0.9 m thick and was mined at Maryhill [6504 6961] and possibly at Cawder Cuilt [5650 7033]. A thinner seam in the same part of the sequence was mined at Springburn [607 680]. The Moor Rock Coal of Giffnock reaches 0.5 m. The South Arthurlie Coal is up to 0.75 m thick and was mined at Barrhead. A maximum thickness of 1.7 m in the Cadder area is attained by the Upper Hirst Coal, mined from Blackhill Colliery [578 714]. In the Darnley area the North Brae or Auchinback Coal (one of the Arden Coals) reaches 1.3 m and was mined near Barrhead. The underlying seatclay was mined as the Darnley Fireclay.

Sandstones There are two kinds of sandstone in the Upper Limestone Formation. The first is fine grained, normally with a gradational base. Most of the thinner sandstones are of this kind. The three thickest of this type are the Bishopbriggs Sandstone, which decreases markedly in all directions from its maximum of 32 m at the type locality (Plate 6b), and the sandstones above the Orchard and Calmy limestones. The second kind is generally medium- or coarse-grained sandstone, with pebbles up to 4 cm across in places and a markedly erosive base. Two such beds, the Upper Dumbreck and Cadgers Loan sandstones, are quite widespread north of the River Clyde. They lie between the Huntershill Cement and Lyoncross limestones, locally removing the former and replacing much of the coal-cyclic sequence normally found between the two limestones. South of the river the two sandstones unite to form the Barrhead Grit which is 55 to 58 m thick in the Darnley area and along its southern margin cuts right down to the Index Limestone. The Giffnock Sandstone is 18 m thick at Giffnock where other sandstones occupy most of the rest of the 50 m thick Lyoncross–Orchard interval. These sandstones are absent in the Darnley area and central Glasgow, where the sequence is only 30 m thick, but reappear north-west of the city, where the same interval is 70 m thick and almost entirely occupied by sandstone, as in Cadder No. 1 Borehole [5879 7187]. The highest sandstone was recorded as medium to coarse grained in Cawder Cuilt No. 1 Borehole [6593 7042] but mostly these sandstones are fine or medium grained, a feature which made the Giffnock Sandstone, like the Bishopbriggs Sandstone, much sought after for building stone.

PALAEOGEOGRAPHY

The Upper Limestone Formation shows marked contrasts in palaeogeography. At least four times, the district (and much of the Midland Valley of Scotland) was inundated by a shallow sea in which thrived the richest faunas known in western Europe, according to Wilson (1967, p.466), except that both goniatites and corals were rare. These marine incursions were thought by Wilson most probably to have come from the east. Several had a considerable duration, the full extent of which is uncertain because the absence of marine fossils in siltstone or sandstone does not prove the absence of marine conditions; for instance, marine or quasimarine conditions may have been continuous during the Index–Huntershill interval. In between these incursions deltaic conditions prevailed; swamp vegetation periodically, but usually more briefly than in the earlier Limestone Coal Formation or the later Coal Measures formations, colonised the lower alluvial plains of the delta top and quasimarine conditions occasionally occurred locally. The major feature, however, was the deposition of sand, some of it coarse and pebbly, probably in river channels cut, in places very deeply, into the deltaic accumulations in the basin whose margins probably lay at some distance from the Glasgow district.

Passage Formation

The Passage Formation where fully developed is over 300 m thick and contains both Namurian and Westphalian A strata, but in this district it is thin and largely unfossiliferous. The formation occurs in south and east Glasgow, where some of the sandstones are exposed on Necropolis Hill [65 60], and near Cadder [58 73] north of Glasgow, where only the lowest beds are preserved unexposed beneath thick Quaternary deposits. Their presence near Cadder is known only from Cadder No. 6 Borehole, which cut more than 40 m of sandstone and fireclay, plus a limestone, 0.35 m thick with 1.5 m of mudstone above it, that may be the Roman Cement Limestone.

Neither the Castlecary Limestone at the top of the underlying Upper Limestone Formation, nor the Lowstone Marine Band at the base of the overlying Lower Coal Measures, has been recognised in south and east Glasgow. The thickness of the Passage Formation therefore is uncertain but appears to be about 85 m, for instance in a borehole [6039 6240] in Govan No. 5 Pit (Forsyth, 1961, p.229). It consists mainly of sandstones, some of which are coarse grained with quartz pebbles up to 2.5 cm across.

Unbedded mudstones, usually mottled red, purple, green and yellow, also occur; some are seatclays and others may merit being called fireclays. Bedded mudstones and coals are rare and thin. The only record of a marine band is in Hampden Park No. 9 Borehole [5889 6152] in which a limestone, 0.25 m thick, and the overlying mudstone yielded a sparse fauna, including productoid brachiopods, which suggests that it belongs to the Passage Formation, probably to No. 0 or No. 1 Marine Band.

The Passage Formation, therefore, is a dominantly fluvial sequence, full of minor unconformities. The depositional basin probably contracted at this time, almost completely excluding the sea, and subsidence was reduced so that the basin was repeatedly filled with fluvial sand, thus minimising the opportunities for coal-forming plant debris to escape destruction below the water table. The basal unconformity cut down into the Upper Limestone Formation to about Plean No. 2 Limestone and repeatedly river channels were cut across the area as sand was alternately deposited and eroded. Away from these channels mud locally and intermittently accumulated on the floodplains.

COAL MEASURES

The coal-bearing, cyclic, fluviodeltaic sedimentary rocks which overlie the Passage Formation have long been known as 'coal measures'. In Scotland they were formerly divided into the Barren Red Measures overlying the Productive Coal Measures, the latter being split into Upper and Lower parts. MacGregor (1960) renamed the units as the Upper (Barren) Coal Measures and the Productive Coal Measures. He divided the latter into the Middle Coal Measures and the Lower Coal Measures. The top of the Middle Coal Measures was defined by the Skipsey's (Aegiranum) Marine Band and the base by the Queenslie (Vanderbeckei) Marine Band. The base of the Lower Coal Measures was drawn at an arbitrary level, taken to be the Lowstone Marine Band (Francis et al., 1970). The base of the Lower Coal Measures as currently in use in Scotland is at a somewhat higher chronostratigraphical level than defined elsewhere. The term Coal Measures (Scotland) was used to reflect the different usage in Scotland. However, the use of the group name, Coal Measures, has been recommended for Scotland as well as in England and Wales; (Table 1; Browne et al., 1996).

Lower Coal Measures

The Lower Coal Measures belong to the Langsettian (Westphalian A), the lowest part of which is either absent or undefinable in the Passage Formation. They occur unexposed in south and east Glasgow, cut off to the southwest by the Dechmont Fault. Dips are generally to the south-east. Only a few borehole records are available; none covers the whole sequence. The stratigraphy (Forsyth, 1979) is still imperfectly known (Figure 13). The thickness is about 100 m. The top is well defined by the base of the Vanderbeckei (Queenslie) Marine Band, but the Lowstone Marine Band at the base has not been found.

STRATIGRAPHY

The sequence may be divided into three parts. The lowest part, up to the Coatbridge Balmoral Coal (that is, to about the top of the Lenisulcata Chronozone) is 20 m thick and consists almost entirely of sandstones and seatearths with a few thin coals and mudstones. It is unfossiliferous except for a band with *Curvirimula*, *Leaia* and *Geisina arcuata*, which probably lies close to the Lowstone Marine Band horizon and is taken as the local base of the Coal Measures.

The middle part, 40 m thick, lies between the Coatbridge Balmoral and Upper Drumgray coals (approximately the Communis Chronozone) and contains more and thicker coals (up to 0.75 m thick), siltstones and mudstones. None of these coals is known to have been worked. Nonmarine bivalves occur above most of the coals, but they tend to be scarce and poorly preserved. There is only one record of the normally abundant assemblage of *Carbonicola pseudorobusta* just below the Lower Drumgray Coal and none of the similarly abundant one above the Upper Drumgray Coal.

The highest part, also about 40 m thick, shows a return to a dominantly sandstone–seatearth sequence, but with several coals up to 0.75 m thick. In the Broad Street and Soho Street area [61 64] the Ladygrange, Bellside and Airdrie Virtuewell coals form a complex of coals and seatclays. The Kiltongue and Airdrie Virtuewell coals were mined at depth from Govan No. 5 Pit and possibly elsewhere.

PALAEOGEOGRAPHY

The dominantly fluvial conditions of the Passage Formation gradually gave way to a broad flat coastal deltaic plain which became fully established in the Middle Coal Measures. During the Lower Coal Measures fluvial influences remained strong, but coal-swamp conditions occurred frequently and at times were particularly persistant. Subsidence was somewhat greater than in the Passage Formation and nonmarine faunas had access to the Glasgow area, but did not flourish there. Only at the base did any approach to marine conditions occur.

Middle Coal Measures

The Middle Coal Measures comprise the whole of the Duckmantian (Westphalian B). They occur in a limited area in south and east Glasgow, bounded to the southwest by the Dechmont Fault. The strata mostly dip to the south-east but in the east there is a reversal to WNW. Surface exposures are few. The stratigraphy is derived from numerous boreholes, mostly put down for urban development, plus a few sections from shafts put down to exploit the thick coals, mining of which has long since ceased. The sequence is 160 m thick (Forsyth and Brand, 1986), which is less than is found in most Scottish coalfields. It contains an unusually high proportion of mud and silt, and rich nonmarine faunas; both features indicate a tendency to return to the quiet backwater conditions of late Dinantian and early Namurian times, in a basin subsiding at a moderate rate. Fluvial sandstones are rare. After the initial marine transgression, represented

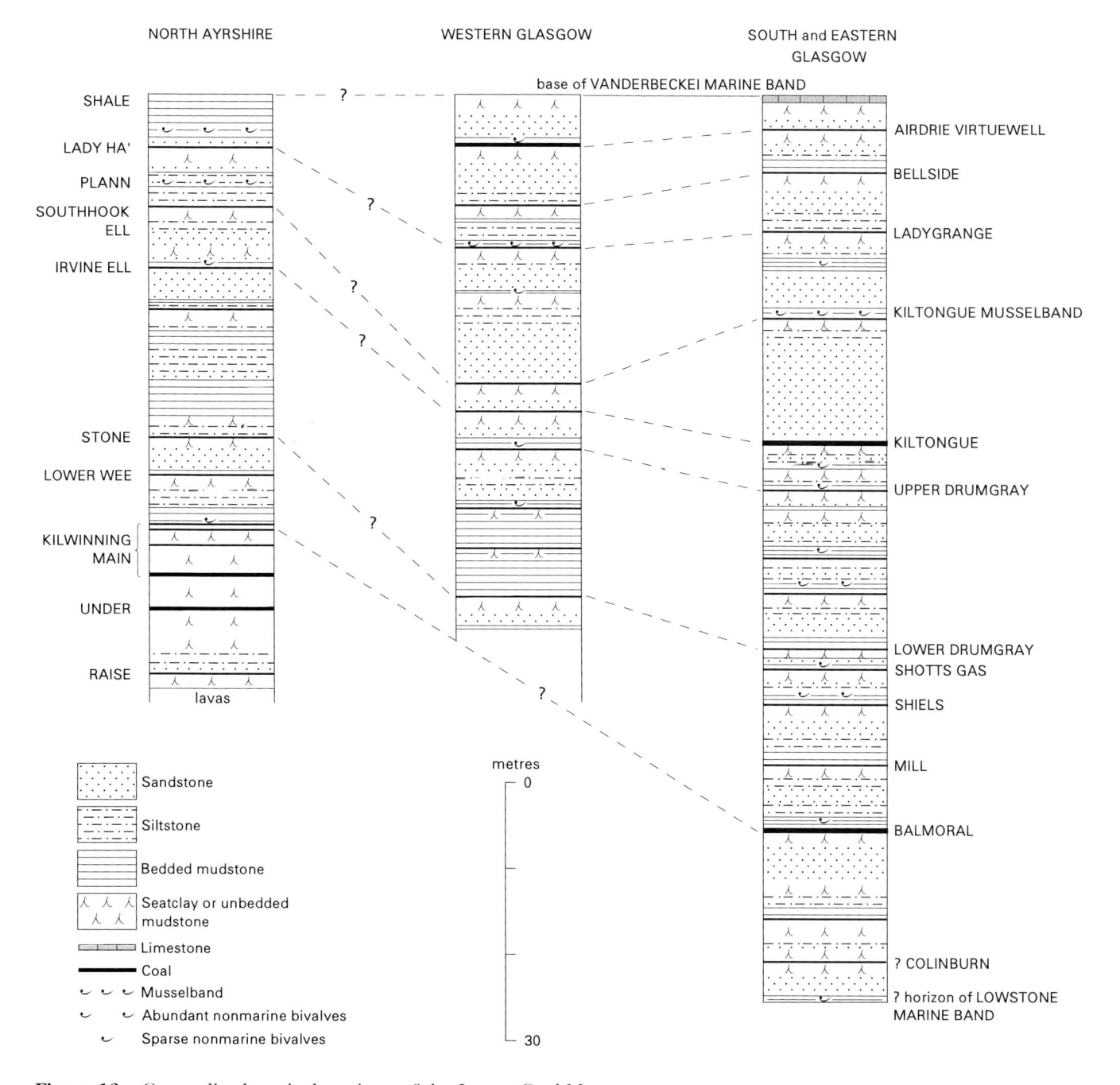

Figure 13 Generalised vertical sections of the Lower Coal Measures.

by the Vanderbeckei (Queenslie) Marine Band (Brand, 1977), there were only two brief recurrences of quasi-marine conditions. The stratigraphy up to the Glasgow Upper Coal is shown in Figure 14. Most of the coals are thinner than they are in the Airdrie district to the east (Forsyth et al., 1996), but the Virgin and Glasgow Upper coals are notable exceptions. The Coatbridge Mussel-band is the most poorly developed of the nonmarine bivalve bands. The Cambuslang Marble lies unusually close to the Glasgow Main Coal but is well developed, with its unusual combination of *Euestheria* and nonmarine bivalves, the latter being notably different from those in the lower bands. The strata above the Glasgow Upper Coal are rather different and less well known. Faunas are sparse, channel sandstones occur and coals are thin; only one very small working is known. The *Euestheria* band which characteristically occurs above the Glasgow Upper Coal in the Airdrie district is poorly developed, especially in Govanhill, where there is a mudstone 10–12 m thick, near the top of which sparse occurrences of *Lingula mytilloides* represent the Glasgow Upper Marine Band. The Drumpark Marine Band is known only as a *Lingula* band in Hamilton Road Route Borehole No. 20 [6093 6287]. It is the only other marine horizon present. The

Figure 14 Vertical section of the Middle Coal Measures up to the Glasgow Upper Coal.

Shettleston Sandstone is over 20 m thick in the Prospecthill Borehole [5999 6184], where it cuts down to 2 m above the Glasgow Upper Coal.

PALAEOGEOGRAPHY

After the major marine incursion that formed the Vanderbeckei Marine Band, sedimentation on the coastal delta plain became re-established. The delta continued to subside and receive large amounts of sand, silt and mud. At regular intervals conditions were established in which first coal-forming vegetation and then nonmarine bivalves flourished.

Upper Coal Measures

The Upper Coal Measures occur in a very limited area in south-east Glasgow. They lie in a broad syncline and are cut off to the south by the Dechmont and Rutherglen faults. Only the lower part (about 100 m thick) is present; it belongs entirely to the Bolsovian (Westphalian C). There are scarcely any exposures and the only borehole of consequence is Prospecthill [5999 6184], where the sequence consists of 3 m of mudstone at the base, containing the Aegiranum (Skipsey's) Marine Band; the Westmuir Sandstone (60 m thick and mainly fine to medium grained but with medium- to coarse-grained bands); and, at the top, 23 m of mainly argillaceous rocks, including a thin coal and containing the Bothwell Bridge Marine Band. Some of the strata are red or purple stained, but the staining is not intense and some beds retain their original pale grey colour. The fauna of the Aegiranum Marine Band at Prospecthill is typical of the western end of the Central Coalfield (Forsyth and Brand, 1986).

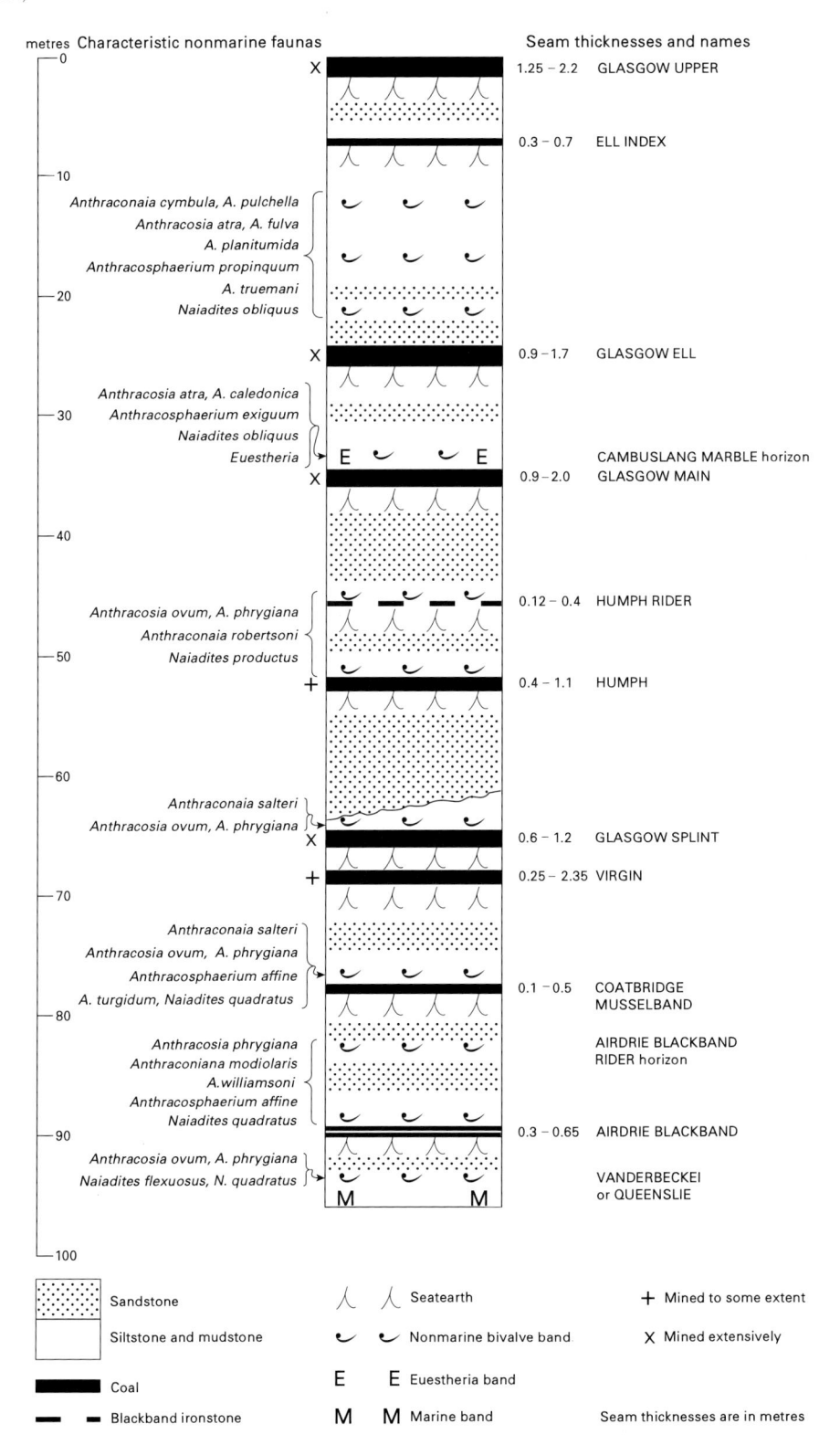

SIX
Carboniferous palaeontology

Comprehensive accounts of the marine faunas of Dinantian and Namurian rocks of the Midland Valley of Scotland have been published by Wilson (1989 and 1967 respectively). Forsyth and Brand (1986) have indicated the character and distribution of nonmarine faunas in rocks of Duckmantian and Bolsovian (Westphalian B and C) age over a similar area. These accounts show the variety of fauna present in rocks of these ages, and also indicate the similarities in the marine faunas throughout the Dinantian and Namurian. In this account emphasis is placed on those characters of faunas that are distinctive, or unique to the district (Plate 7). However, at many horizons it is more often the relative abundance of individual species in collections that enables a particular horizon to be identified.

BALLAGAN FORMATION

In the district exposures of this formation have yielded few fossils other than *Spirorbis* sp. and ostracods. In both the Barnhill [4269 7571] and Loch Humphrey [4592 7555] boreholes, strata assigned to the formation contain scattered specimens of '*Estheria*' sp. and ostracods. In Murroch Glen [4192 7845] in the Greenock district, a *Lingula* Band has been recorded from strata assigned to the formation. The only other record of *Lingula* from strata of this formation is from near the Heads of Ayr [2964 1875].

CLYDE PLATEAU VOLCANIC FORMATION

Several exposures of intercalated ashy sedimentary rocks of the Greenside Volcaniclastic Member occur and the majority of these contain plants. Scott et al. (1984, pp.317–318) have listed the flora from a locality near Loch Humphrey [468 755] and similar floras occur elsewhere, usually associated with layers containing fish remains. At Rigangower Quarry [4400 7512], a bed near the base of the sequence contains plant debris, an estheriid and fish fragments.

LAWMUIR FORMATION

Although marine horizons at a slightly lower level are known, the Dykebar Limestone (or Balmore Marine Band in the north) is the lowest marine horizon which may be traced over most of the district. Even then it is absent in the south-west limb of the Howwood Syncline, and was not recognised in the Inchinnan area. *Spirifer* of the *crassus* group is characteristic of the faunas from this horizon south of the River Clyde, and at least as far as the Lawmuir Borehole [5183 7310] north of the river, but it is absent from faunas which have been collected near Milngavie. Elsewhere in Scotland the correlatives of this horizon contain examples of *Posidonia becheri* and the bed is thus thought to be of P_{1b} or P_{1c} age. *Latiproductus* cf. *latissimus* is present in abundance in the succeeding Hollybush Limestone over a similar area, but is absent from faunas east of Milngavie in this district. The fauna of the Blackbyre Limestone does not contain any particularly characteristic species, nor are there any marked geographical changes. Wilson (1989, p.104) drew attention to the large number of brachiopod species occurring at this horizon. These include *Alitaria* spp., *Antiquatonia hindi*, *A. insculpta*, *A. sulcata*, *Avonia youngiana*, *Beecheria hastata*, *Buxtonia* sp., *Composita ambigua*, *Crurithyris urii*, *Echinoconchus elegans*, *E. punctatus*, *Eomarginifera* spp., *Latiproductus* sp., *Lingula* spp., *Orbiculoidea* sp., *Phricodothyris lineata*, *Pleuropugnoides* sp., *Productus concinnus*, *Pugilis pugilis*, *Rhipidomella michelini*, *Rugosochonetes* sp., *Schizophoria resupinata*, *Spirifer bisulcatus* group, *S. crassus/striatus* group and *Tornquistia polita*.

LOWER LIMESTONE FORMATION

Wilson (1989, fig. 8, pp.110, 112–113) has shown that there are marked changes in faunas to the north and north-west of Glasgow at the horizons of the Hurlet and Blackhall limestones and suggested that these were related to environmental differences resulting from the presence of a river system supplying sediment and fresh water from this direction. The changes noted in the limestone faunas of the Lawmuir Formation may have had a similar origin. Specimens of the ammonoid *Dimorphoceras marioni* from the Neilson Shell Bed above the Blackhall Limestone indicate a Brigantian (P_2 Zone) age (Currie, 1954, p.534). The type specimen is from the *Sudeticeras splendens* Zone (P_{2b}) of the Pennines (Moore, 1939, pp.120–121). The overlying Hosie limestones contain distinctive faunas, listed by Wilson (1989). The presence of rare specimens of *Posidonia membranacea* at this level is of interest, suggesting links to beds of Brigantian (P_2 Zone) age in the Pennine area, where the species is much more common.

LIMESTONE COAL FORMATION

The mudstones above, and just below, the Top Hosie Limestone contain rare specimens of the goniatite *Cravenoceras scoticum* which resembles *Cravenoceras leion*, the incoming of which marks the base of the Namurian in Europe. Despite this, in this district, the base of the Namurian (Pendleian Stage, E_1 Zone) is drawn conven-

44 SIX CARBONIFEROUS PALAEONTOLOGY

Plate 7 Selected Carboniferous fossils.

a. *Echinoconchus punctatus* (GSE 8486) × 1. Lower Limestone Formation, Hurlet Limestone, Corseford Limeworks [4110 6086].
b. *Actinopteria regularis* juv. (GSE 11850) × 2. Upper Limestone Formation, Calmy Limestone, Darnley Quarry [5254 5978].
c. *Leptagonia smithi* (GSE 12813) × 1. Upper Limestone Formation, Orchard Limestone, Darnley No. 3 Borehole at 42.32 m [5386 5878].
d. *Rayonnoceras windmorense* (GSE 4627) × 1.5. Upper Limestone Formation, Orchard Limestone, Orchard Quarry [5615 5875].
e. *Caneyella membranacea* (GSE 12144) × 1.5. Lower Limestone Formation, Mid Hosie Limestone, Milngavie No. 1 Borehole at 103.02 m [55017335].
f. *Posidonia corrugata gigantea* (GSE 12146) ×1.5. Lower Limestone Formation, Neilson Shell Bed, Milngavie No. 2 Borehole at 87.17 m [5392 7357].
g. *Pugnax* cf. *pugnus* (GSE 8525) × 1.5. Upper Limestone Formation, Calmy Limestone, quarry near Davieland [5561 5824].
h. *Linoproductus* cf. *concinniformis* (GSE 11385) × 1.5. Upper Limestone Formation, Orchard Limestone, Thornliebank [5461 5910].

tionally at the top of the Top Hosie Limestone. Strata subsequent to this marine episode contain two principal marine bands, the Johnstone Shell Bed and the Black Metals Marine Band. The first of these contains the more varied fauna, and is also characterised by the abundance of species of *Productus*, whilst in the second the commoner productoid is *Buxtonia* sp. The fauna of these marine bands shows no marked variation when traced from south to north. Forsyth (1980) has shown that the various *Lingula* bands in the upper part of the formation, above the Black Metals Marine Band, do show marked geographical changes which might be taken to indicate the direction of the marine incursion. Thus, for example, the *Lingula* band above the Shale Coal dies out eastwards towards the Stirling area, as does that above the King Coal, whilst that above the Fourteen Inch

Under Coal, found in the east around Bedlay, in the Airdrie district, has yet to be found within the Glasgow district. The nonmarine faunas in the group are restricted in variety.

The dominant species of *Lingula* in the *Lingula* bands is *L. squamiformis*. This form is particularly abundant in the lower horizons of the formation, where individuals are also, in general, larger than those occurring higher in the formation. The species is also abundant in the persistent band above the Bo'ness Splint Coal, which has occasionally also yielded *Orbiculoidea* sp. or indeterminate marine bivalves.

Paracarbonicola is found in the upper part of the mudstones above the Top Hosie Limestone, in abundance in the Linwood Shell Bed and in the upper part of the Johnstone Shell Bed. In the Linwood Shell Bed the genus may be found associated with *Lingula* sp. and rare marine bivalves. Species of *Curvirimula* and *Naiadites* appear to be mutually exclusive, and the latter genus appears to have flourished when the more canneloid mudstones were laid down such as that associated with the Lower Possil Ironstone. Some of the ironstones, notably the King Blackband Ironstone, contain dwarfed nonmarine bivalves and ostracods. This ironstone also contains shards of *Lingula* sp., suggesting winnowing of the shell material or current transport of finely broken material.

The famous Fossil Grove at Victoria Park [540 672] consists of casts of the lower rooted portions of *Lepidodendron* trees preserved in sandstone (Plate 8). The horizon occurs in the lower part of the formation, possibly above the Johnstone Clayband Ironstone.

UPPER LIMESTONE FORMATION

Revision of the few zonally significant goniaties by Ramsbottom (1977) has shown the presence of *Tumulites pseudobilinguis* in the mudstones above the Index Limestone indicating a Pendleian age (E_{1b2}). *Eumorphoceras grassingtonense* from the Orchard Limestone at Orchard [565 594] indicates an Arnsbergian age (E_{2a2}), and '*Cravenoceras*' *gairense* and *Eumorphoceras ferrimontanum* from the Calmy Limestone an Arnsbergian age (E_{2a2}). These occurrences help to define the limits of the Namurian stages within the Upper Limestone Formation.

The Index Limestone is characterized by the presence of *Latiproductus* cf. *latissimus* and algae. The marine fauna in the overlying mudstones includes *Serpuloides carbonarius*, *Euphemites* sp., *Meekospira* sp., *Retispira* sp., *Palaeoneilo luciniformis*, *Posidonia corrugata* and *Streblopteria ornata*. The mudstones associated with the succeeding Huntershill Cement Limestone contain a similar but impoverished fauna. The fauna of the Lyoncross Limestone includes the bivalve *Streblopteria ornata* which marks its last occurrence in the Scottish Namurian. The Orchard Limestone contains the most varied fauna in the formation; in some features this fauna resembles that of the Neilson Shell Bed. The brachiopod *Antiquatonia costata* is almost entirely confined to this horizon in the Namurian; the gastropod *Straparollus carbonarius* is of frequent occurrence while the presence of numbers of *Palaeoneilo mansoni* and *Posidonia corrugata* in the mudstones is noteworthy. The mudstones associated with the Calmy Limestone contain distinctive faunas. *Edmondia*

Plate 8 Fossil Grove, natural casts of lower rooted portions of *Lepidodendron* trees (Limestone Coal Formation) (D 1536).

punctatella occurs in a thin carbonaceous mudstone below the limestone throughout the district, and in the mudstones above the limestone the brachiopods *Pugnax* cf. *pugnus* and *Sinuatella* cf. *sinuata* occur only at this horizon. *Serpuloides carbonarius* and *Euphemites ardenensis* also occur as distinctive elements in the fauna (Wilson, 1967, p.459). However, there is insufficient evidence in this area to indicate any change in faunas in a geographical sense.

PASSAGE FORMATION

Little is known of faunas in this formation in the district.

LOWER COAL MEASURES

Recent drilling in the upper part of the Lower Coal Measures in western Glasgow has shown the presence of a variable sequence of mainly thin coals with a number of fossiliferous horizons. Some of these faunas are unlike those occurring at similar levels in other parts of the Central Coalfield. Forsyth (1979) discussed the known stratigraphy and faunal distribution in eastern and central Glasgow.

A feature of the faunas of the Lower Coal Measures in eastern and central Glasgow is the persistence of the band containing *Planolites* aff. *ophthalmoides* above the Mill Coal. This form is larger than *Planolites ophthalmoides* and may be a distinctive form and not necessarily representative of an approach to marine conditions.

In western Glasgow [576 625] strata as low as the Lower Drumgray Coal have been proved. Here a poor mussel-band above the upper of two seams between the Lower Drumgray and the Upper Drumgray coals contains *Curvirimula* sp., contrasting with the development farther east where *Leaia* and other estheriids are also present. The Upper Drumgray Musselband is not well developed, contrasting with other areas of the Central Coalfield. Similarly, the Kiltongue Musselband, elsewhere a very recognisable and characteristic horizon, is poorly developed or not recognisable in the district. A distinctive fauna of *Anthracosia regularis*, *Carbonicola oslancis*, species of *Naiadites* and abundant specimens of *Geisina arcuata* occurs above the correlative of the Ladygrange Coal, in contrast to localities farther east, where this horizon is usually barren. Faunally and lithologically the bed is similar to some occurrences of the bed above the Lady Ha' Coal of North Ayrshire. So far no unequivocal correlation exists with the Lower Coal Measures of North Ayrshire. A poor fauna with *Anthracosia regularis* and species of *Naiadites* is present above the Airdrie Virtuewell Coal, but only where it is not removed at the base of the overlying erosive sandstone. Figure 14 illustrates the differences between the sequence in Glasgow and those in adjacent districts.

MIDDLE COAL MEASURES

The fauna of the Vanderbeckei Marine Band and its distribution was discussed by Brand (1977). Discoveries since that date have served to confirm the pattern of distribution of faunal facies in which the productoid facies is present over much of the outcrop in this district. The palaeoecological inferences of this occurrence and those in adjacent districts to the east were discussed by Brand who suggested the influence of a river system to the north or north-west of the area. Such a system would occupy a similar location to the one that Wilson (1989) thought existed during the period in which the Lower Limestone Formation was deposited. Faunas at horizons above this level are similar to those known elsewhere in the Central Coalfield, although the canneloid mudstone development above the Humph Rider Coal, containing specimens of *Anthraconaia* which approach *A. polita* in outline, is unique to the district.

The fauna of the Cambuslang Marble is characterised by the incoming of a varied assemblage including, amongst other species, *Anthraconaia cymbula*, *A. librata*, *Anthracosia aquilina* Trueman and Weir *non* Sowerby, *A. atra*, *Anthracosphaerium* cf. *truemani* and *Naiadites* of the *obliquus* group. The presence of *Euestheria* at this horizon in association with this assemblage is distinctive. The *Euestheria* band in the roof of the Glasgow Upper Coal, normally well developed in the adjacent Airdrie district, is here poorly developed. At Govanhill [593 625] sparse occurrences of *Lingula mytilloides* near the top of the mudstones above the Glasgow Upper Coal represent the Glasgow Upper Marine Band (this may be correlated with the Two Foot Marine Band of the Pennine area). A more varied marine fauna including *Myalina*?, pectenids and *Holinella* sp. occurs above the coal in a restricted area around the Clyde Ironworks [645 623] in the Airdrie district but the marine band appears to be absent between the two occurrences.

UPPER COAL MEASURES

The development of a *Lingula*–foraminifera facies at the horizon of the Aegiranum Marine Band has been used as an indicator for shoaling of the marine environment westwards from Hamilton and Coatbridge (Forsyth and Brand 1986, p.17). In the Prospecthill Borehole the Aegiranum Marine Band contains *Planolites ophthalmoides*, foraminifera, sponge spicules, *Lingula* sp., *Orbiculoidea* sp., the marine ostracod *Holinella* and conodonts. A similar but poorer fauna was obtained in the Govanhill No. 27 Borehole [5931 6252]. The fauna from the overlying Bothwell Bridge Marine Band consists of *Planolites ophthalmoides*, foraminifera, *Lingula*? and ostracods.

MICROFLORA

The absence of distinctive macrofaunal horizons in the lower part of the Carboniferous has stimulated investigation in the miospore content of these sedimentary rocks. There appears to be little controversy concerning the miospore zonation of the Ballagan Formation, most workers recording the presence of forms characteristic of the CM Zone of Courceyan age. However, Owens (*in*

Paterson and Hall 1986, p.8) has recorded miospores characteristic of the TC Zone of late Holkerian/early Asbian age in a sample from strata in this formation in the Airdrie district. Above this level, sedimentary rocks intercalated in the Clyde Plateau Volcanic Formation have yielded a variety of results. Scott et al. (1984, pp.314–318) have described miospore assemblages belonging to the CM and Pu zones from one 25 m thick intercalation (Greenside Volcaniclastic Member) at Loch Humphrey. Rocks from the upper part of this section contain a miospore assemblage indicating a Pu Zone age with some elements of the TC Zone assemblage (Scott et al., 1984, p.317). Miospore floras from the Lawmuir Formation indicate the presence of the NM Zone of Asbian age in the lower part, whilst strata above the Dykebar Limestone yield miospores characteristic of the VF Zone of Brigantian age. Since the work of Butterworth and Williams (1958) and Smith and Butterworth (1967) little new work has been published on the miospore zonation of younger Carboniferous rocks in the district. However, Neves et al. (1972) have established a miospore zonation for the Lower Carboniferous of Scotland which is applicable to rocks in the district.

Theses by Dean (1987) and Hutton (1965) have examined the conodont and foraminiferal assemblages in the Lower and Upper Limestone formations, but little of their work was carried out on material from this district.

SEVEN
Late- and post-Carboniferous intrusive rocks

QUARTZ-DOLERITE DYKES

Dykes belonging to the Midland Valley quartz-dolerite suite, with an approximately west–east trend, occur in the northern part of the district. They are late Westphalian to early Permian in age (295–290 Ma, de Souza, 1979). None can be traced continuously across the whole district but some extend unbroken for at least 7 km. The Lenzie–Torphichen dyke which starts north of Blackhill [572 708] extends eastwards for 40 km across the Airdrie district into the Falkirk district. The dykes cut rocks of Upper Devonian to Namurian age, and are up to 30 m wide. The Campsie, Mugdock and Milngavie dykes cut only Lower Carboniferous rocks and are known almost entirely from numerous, mostly prominent exposures, as are the two that cut the Lawmuir Formation near Bishopton. The others are known from mining information. Two such dykes in the Linwood area that were previously mapped as continuous quartz-dolerites, are now regarded as intermittently occurring basaltic intrusions of uncertain affiliation.

Petrographically the quartz-dolerites are olivine-free subophitic dolerites and basalts, consisting of basic plagioclase, augite (usually with some hypersthene) and abundant titaniferous magnetite, with a mesostasis of micropegmatite or (usually devitrified) glass.

The Midland Valley Sill is unknown in this district.

ALKALI DOLERITE SILLS

A sill-complex consisting of several leaves of olivine-dolerite of teschenitic affinities (formerly called teschenite), extends from Kilbarchan (Greenock district) through Johnstone and Paisley to Glasgow. Radiometric dates of 270 Ma have been obtained using biotite from Cathcart [58 60] and hornblende from Barshaw [506 642] and of 273 Ma using biotite from the Necropolis [605 654] (de Souza, 1979). They form a good cluster and indicate a lower to mid-Permian age. The sill-complex exceeds 50 m in thickness south of Johnstone, where it is intruded into the Lawmuir Formation just above the Quarrelton Thick Coal. A sill about 12 m thick lies at a similar horizon south of Elderslie [445 632], but to the east there is a gap of 3.5 km to central Paisley, where a sill lies at the top of the formation. Other sills occur, for example in the railway cutting at Arkleston [500 650], beside the White Cart Water [498 629] and north of Hurlet [515 609]. The sill formerly exposed at Barshaw is now thought to lie just below the Blackhall Limestone. It includes the famous bekinkinite of Barshaw, which consists largely of purple augite and brown hornblende in about equal proportions, with some olivine pseudomorphs plus a little plagioclase, magnetite, apatite and a relatively abundant matrix of chlorite, analcime, etc. (Clough et al., 1925).

East of Barshaw dolerite has been recorded in a number of boreholes and was formerly visible at Cardonald [5242 6418]. The next occurrences are to the north, in the vicinity of the River Clyde east of Renfrew. Data about them comes from boreholes, one of which proved 45 m of dolerite in several leaves, a temporary exposure [5166 6816] by the river east of Yoker and a more permanent one at the Fossil Grove [538 673]. These sills are intruded into the upper part of the Lower Limestone Formation and the lower part of the Limestone Coal Formation. To the east of them there is another gap of 6 km in the surface exposures to those on the Necropolis Hill [60 65], where the dolerite is probably in the Passage Formation and transgresses upwards into Lower Coal Measures at the eastern margin of the district. Dolerite has also been encountered in boreholes and temporary exposures in central Glasgow, up to 1.5 km west of the Necropolis, where the sill is probably in the Passage Formation or high in the Upper Limestone Formation.

Three boreholes in northern Glasgow, namely Gartnavel No. 3 [5578 6781], Maryhill [5718 6856] and Colston Road [5945 6913], all encountered dolerite in the Lower Limestone Formation or basal Limestone Coal Formation and proved thicknesses of 37 m (Colston Road), 40 m (Gartnavel) and 80 m in several leaves, with a main leaf of 58 m (Maryhill). In addition, thinner sills of unclassed dolerite have been encountered in other boreholes in north-west Glasgow. Alkali olivine-dolerite sills also occur in southern Glasgow at Cathcart, where they are intruded into the upper part of the Limestone Coal Formation. Some of the thinner leaves, mostly carbonated and bleached to form 'white trap', are exposed in the banks of the White Cart Water [582 602] south of Cathcart and were also recorded in several boreholes to the west. A fresher dolerite, at least 10 m thick, occurs beside Cathcart Castle [586 601] and is near the top of Limestone Coal Formation. Aitkenhead No. 4 Borehole [5981 6036] proved another sill 70 m thick, also in the upper part of the formation, which is the thickest single sill known in the whole complex. Another sill, situated farther south-east and therefore probably intruded lower in the formation is exposed beside Castlemilk House [610 596] and was encountered in boreholes, shafts and a sewer tunnel about 1 km to the south-west, where a picritic layer was noted. The total thickness of the alkali dolerite sills in the Cathcart–Castlemilk area clearly exceeds 80 m, and is the maximum amount known in the sill-complex.

The two main minerals in these alkali olivine-dolerite sills are purple-brown augite and calcic plagioclase; their relation vary from hypidiomorphic to ophitic. Pseudomorphs after olivine are common. Biotite, titaniferous magnetite, apatite and analcime are common accessories, but the analcime is insufficiently abundant and the olivine too abundant for these rocks to be classed as teschenites. Brown hornblende and nepheline occur in the Cathcart sills. The Necropolis and adjacent sills have two types of segregation veins. The blue ones are fine-grained olivine-basalts; the pink ones have conspicuous needles of brown hornblende in a feldspathic base which includes large zoned crystals of alkali feldspar, many small feldspar needles, analcime and pseudomorphs after nepheline.

ALKALI DOLERITE DYKES

A dyke of olivine-basalt, 2.5 m wide, has been traced WSW–ENE for 1.5 km in surface outcrops and mine-workings in northern Glasgow. It cuts Limestone Coal and Upper Limestone formation strata and is characterised by abundant fresh olivine. In the Beith–Barrhead Hills, a 3 m-wide WNW–ESE-trending dyke and a 2 m-wide NE–SW offshoot, both of very fresh ophitic olivine-dolerite, cut lavas of the Clyde Plateau Volcanic Formation in the Old Patrick Water [435 601]. The mineralogy and texture of these dykes resembles closely that of the alkali olivine-dolerite sills, with which they are here associated, although their very fresh nature and in particular the fresh olivine, has led previously to such dykes being thought to be of Tertiary age. However, the district lies to the north-east of the sharply defined north-east limit of the Tertiary Mull regional swarm (Cameron and Stephenson, 1985; Paterson et al., 1990) and hence association of these fresh olivine-dolerite dykes with the Permian magmatism seems more likely.

MILNGAVIE SILLS

The Milngavie dolerite sills crop out northwards from the Milngavie–Kilsyth Fault to Mugdock [558 771] and from Milngavie in the west to beyond the Linn of Baldernock [590 757]. They are intruded into the Clyde Plateau Volcanic, the Lawmuir and the Lower Limestone formations and are therefore post-Dinantian in age, but no radiometric dating has been done. Exposures are numerous and mostly prominent. Individual sills are up to 30 m thick. Where they come into contact with the dark grey mudstones of the Lower Limestone Formation or with the Hurlet Coal (as at the Linn of Baldernock) they are extensively carbonated and bleached. Similar sills occur at depth south of the Milngavie–Kilsyth Fault. In Milngavie No. 6 Borehole [5634 7384] a sill 24.2 m thick is intruded just below the Blackhall Limestone and in Milngavie No. 5 Borehole [5735 7285] a sill 19.6 m thick occurs at the Hurlet Coal. The Milngavie sills are hypidiomorphic olivine-free dolerites with abundant basic plagioclase feldspar, pale brown, locally purplish augite and skeletal titaniferous magnetite. Small amounts of biotite and brown hornblende occur. The mesostasis consists of secondary chlorite and carbonate. No clear relationship can be established to either the quartz-dolerites or the alkali olivine-dolerites, but the nature of the mafic minerals suggests an affinity to the latter.

EIGHT
Structure

REGIONAL CONTEXT AND IMPLICATIONS

The district lies near the centre of the Midland Valley of Scotland, an ancient rift valley bounded in the north by the Highland Boundary Fault and in the south by the Southern Upland Fault. The graben structure was developed by the early Devonian (Bluck, 1978) in a zone of crustal weakness inherited from the Lower Palaeozoic. Analyses of structural patterns, facies distribution of the sedimentary rocks, and the nature of the volcanism, have led to the proposal of a wide variety of stress systems and their origins, which are thought to have controlled the development of the Midland Valley of Scotland in the Devonian and Carboniferous. Recent suggestions include east–west extension (Russell, 1971; Haszeldine, 1988; Stedman, 1988), north–south extension (Leeder, 1982; Leeder and McMahon, 1988) and NW–SE extension with superimposed right-lateral strike slip (Dewey 1982; Read 1988). However, several authors, such as Kennedy (1958) and Read (1988), have pointed out that different stress systems appear to have been operative at different times during the late Devonian and Carboniferous.

The oldest strata seen at the surface in the Glasgow district are Lower Devonian sandstones which were laid down on the broad floodplain of a mature river system with a generally westward axial drainage. During the mid-Devonian, uplift and widespread earth movements culminated in the formation of the NE–SW-trending Strathmore Syncline and the reversal of the regional palaeoslope. After this major break, Upper Devonian sandstones were deposited by eastward-flowing rivers. These fluvial sandstones become gradually finer upwards and are interbedded with sandstones of aeolian origin. The lowest Inverclyde Group sedimentary rocks, of late Devonian–early Carboniferous age, contain extra-basinal pebbles (quartz) indicating some rejuvenation of the hinterland at this time. However, overall tectonic conditions were stable and the alluvial plain on which the sediments were deposited gradually became marginally marine and subject to fluctuating salinity and periodic desiccation. Towards the end of Inverclyde Group times, a return to fluvial sedimentation indicates rejuvenation and further uplift of the source area.

Subsequently, during the mid-Dinantian, a major unconformity occurred, the effects of which are seen along the south margin of the Campsie Fells and along the north margin of the Gleniffer Braes (Figure 15). This unconformity has been related to magmatic updoming prior to eruption of the Clyde Plateau Volcanic Formation lavas (Forsyth et al., 1996) and has a Caledonoid trend. The updoming and erosion were followed by the eruption of the Clyde Plateau Volcanic Formation lavas. These are alkali basalts typical of the volcanic rocks found throughout the world in intra-plate rift systems associated with tensional stress regimes. Some of these volcanic rocks were erupted from a series of linear vent systems aligned ENE–WSW (Craig, 1980) but there are also less clearly defined groupings of vents with a NW–SE trend in the Bowling area and near Blanefield. The Bowling group of vents is at the north-west end of a south-east-trending belt of faults which joins the Dechmont Fault, a complex zone which continues to the south-east where it may have partly controlled late Dinantian and early Silesian sedimentation. The ENE–WSW linear vents are thought to be strongly suggestive of NNW–SSE tension or possibly transtension (Read, 1989). This would not necessarily be negated by the presence of NW–SE-trending structures. However, in the apparent absence of a hot-spot trail in the Midland Valley, Smedley (1986) suggested that passive rifting with gentle upwelling of magma occurred in a region already weakened by Caledonian tectonism. If this is the case, it seems possible that the orientation of the linear vent systems may have been inherited and consequently the apparent trend of the regional tensional stress system at this time may be misleading.

After volcanism ceased, erosion of the lava blocks took place and detritus was deposited locally along the margins while fluvial sediments were being deposited elsewhere. The occurrence of the Douglas Muir Quartz-Conglomerate north of Glasgow indicates uplift of a hinterland to the north-west, where possibly Lower or Upper Devonian conglomerates were being eroded. Subsequently, cyclic sedimentation with marker bands became established and continued throughout the late Dinantian and Silesian. The marker bands potentially allow facies analyses of stratigraphical intervals to be made but, since exposures of these strata are very limited, data are largely from commercial boreholes and consequently heavily weighted to strata of economic interest. During the late Dinantian the cyclic sediments of the Lower Limestone Formation, which in the Glasgow district are mudstone-dominated, indicate repeated subsidence of a fluviodeltaic environment below a shallow sea. Within the district, the sediments are thickest in an ENE-trending basin, the Kilsyth Trough, which was developed in and to the north-east of Glasgow (Browne et al., 1985, fig. 13). They thin rapidly towards the lava blocks lying to the north-west and south-west but in the south-west the controlling factor may have been the Dechmont Fault. A similar pattern is seen in Ayrshire, but to the east of the present district the Kincardine and the west and east Fife basins are developed with a north–south trend. During the Namurian, the Limestone Coal Formation strata of the Glasgow district were deposited in a similar pattern, mainly in a deltaic environment, with frequent quasimarine incursions. On the bases of facies and thickness distribution of

Figure 15 Location of the mid-Dinantian unconformity.

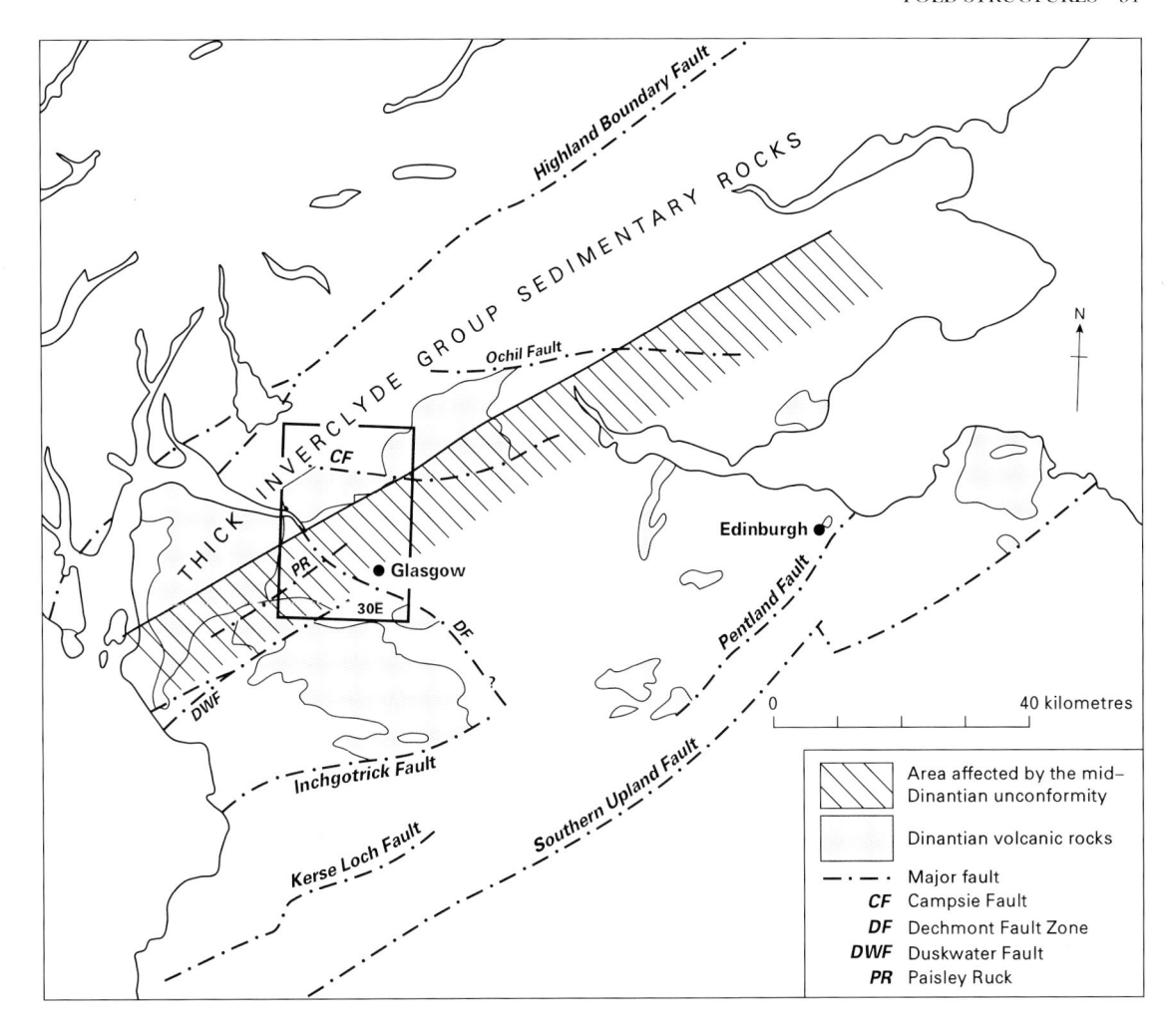

the Limestone Coal Formation sediments, Stedman (1988) suggested that the stress regime was one of east–west tension which in the south-west of the Midland Valley reactivated the Caledonoid grain. Read (1988) on the other hand, proposed that dextral shear was operating during this period. In most of the Glasgow district the succeeding fluviodeltaic and marine sediments of the Upper Limestone Formation show a gradual thinning to the south-west, although their thick development in the north-east appears to be partly an extension of that occurring in the Kilsyth Trough. The dominantly fluvial conditions with short-lived marine incursions which followed during the deposition of the Passage Formation show a marked change in facies and sediment thickness. Many minor unconformities occur reflecting tectonic instability and the overall thickness gradually increases to the north-east of the district. In the central part of the Midland Valley north–south folds replaced ENE–WSW lineaments as the dominant controls on sedimentation (Read, 1989). The overlying fluviodeltaic Lower and Middle Coal Measures of Westphalian age are thickest in the east of the district and thin very gradually westwards reflecting a continuing north–south control. The intrusion of the east–west-trending quartz-dolerite dykes commonly along but later than faults, and now generally accepted to be of late Carboniferous–early Permian age, indicates an apparent north–south tensional phase. Read (1988) concluded that all these Silesian structural features are compatible with right-lateral strike slip superimposed on thermal subsidence. Though the amount may have been relatively small up to the end of the Namurian, it may have been greater at the end of the Carboniferous (Read, 1989).

FOLD STRUCTURES

In the northern part of the district the effects of folding are not apparent, though the Lower Devonian strata lie in or near the axial zone of the Strathmore Syncline, a major fold which runs across the Midland Valley. South of the Milngavie–Kilsyth Fault (Figure 16) the strata are affected by several rather impersistent and in some cases little known folds that mostly have a generally NE–SW trend, and are much dissected by the numerous faults. North-west of the Paisley Ruck a syncline can be traced from Portnauld [492 685] to the Linwood Basin and WSW as far as Johnstone [45 64]. The Paisley Ruck itself is reported (Carruthers *in* Clough et al., 1925, fig. 19, p.201) to pass north-eastwards into a sharp NNE–SSW anticline (with the limbs dipping at up to 45°), which bisects a larger NNE–SSW half basin [53 70], cut off to the north by a fault. The evidence for the existence of this anticline is, however, largely derived from a record

52 EIGHT STRUCTURE

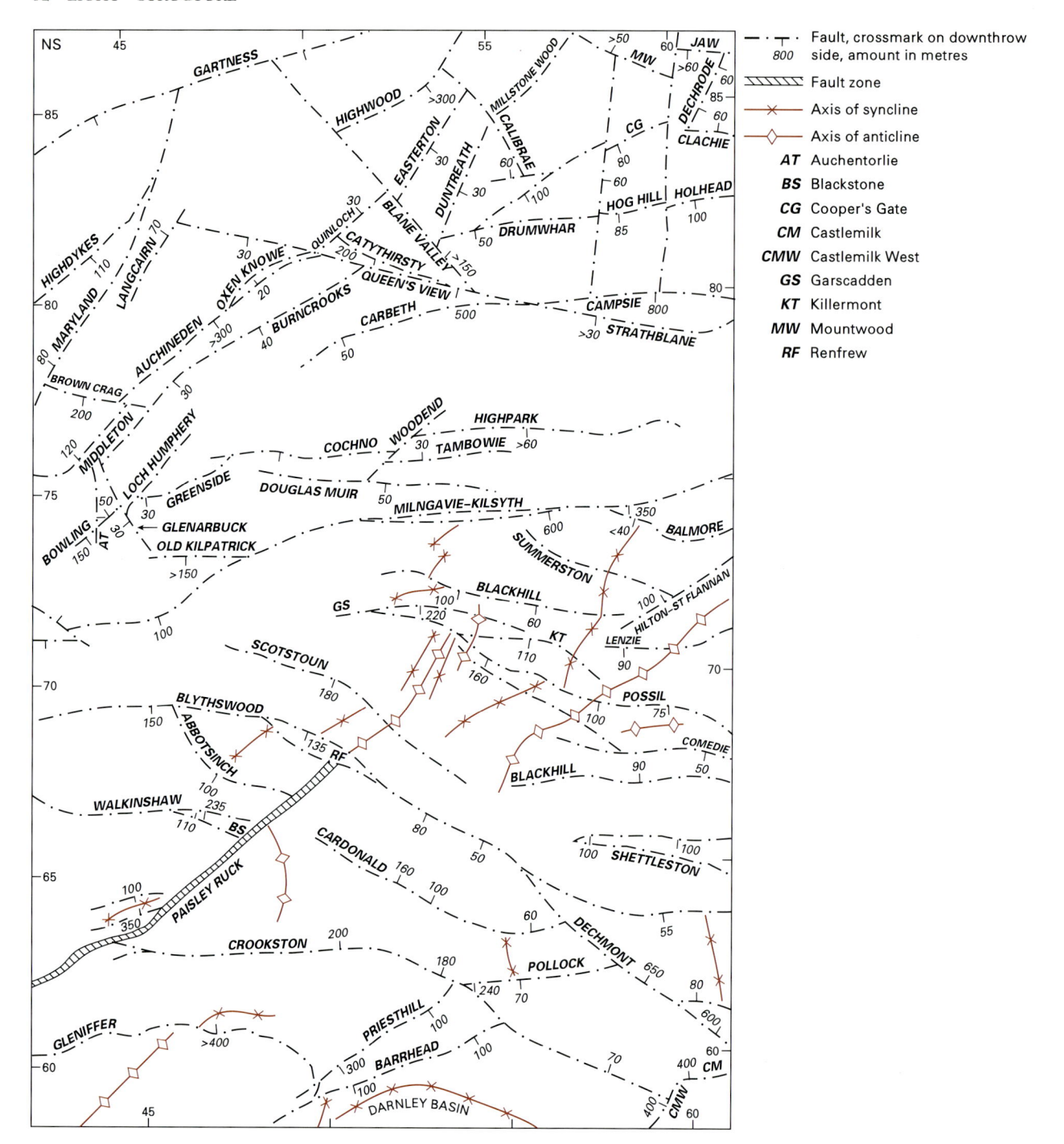

Figure 16 Generalised structure of the Glasgow district showing the principal faults and folds.

of a stone mine [529 697] on an old mine plan (Hinxman et al., 1920, fig. 3, pp.45–46) that is no longer available. Little other information is available and the structure in that area may be quite different.

South-east of the Paisley Ruck the general dip is to the south-east but the structure is locally complicated by minor folds such as the anticline affecting the Quarrelton Thick Coal at Johnstone Castle [432 620], a small basin [461 629] round which the Hollybush Limestone crops out, the small but sharply folded basin at Nethercraigs [466 610] and ill-defined folds affecting the Hurlet Limestone in central Paisley. In southern Glasgow the main fold is the west-east elongated Darnley Basin from Barrhead to Cathcart, with a north-south syncline at Lochinch [553 628] and a half basin in the Upper Coal Measures [60 62]. The only fold affecting the Clyde Plateau Volcanic Formation is an open anticline with a NE-trending axis which occurs in the Beith–Barrhead Hills.

North of the River Clyde the major structure is the NE–SW Blackhill Syncline which can be traced from

Bardowie [584 735] as far as western Glasgow [53 68]. The axis plunges to the north-east so that Passage Formation strata are preserved at that end whereas the Limestone Coal Formation crops out at the south-west end; dips of the limbs reach 30° in the central part. The anticline to the east is less well defined but can be traced from Cadder [615 723] to Partick [553 676], with a subsidiary W–E anticline in northern Glasgow [58 68–59 68].

FAULTS

The more important of the many faults occurring in the district are shown in Figure 16, together with selected throws. In the northern part of the district, the Clyde Plateau Volcanic Formation and older strata are affected by block faulting reflecting their massive largely homogeneous nature. The dominant trends are ENE and NE, though NW, north–south and WNW also occur. In the southern part of the Kilpatrick Hills east–west is the main trend. NE trends, e.g. the Dumbarton–Fintry line, and NW trends e.g. the Blane Valley and Dechmont faults, appear to have controlled the locations of the vents from which the lavas of the Clyde Plateau Volcanic Formation were erupted. The Dechmont Fault may have been active intermittently throughout the Carboniferous and together with the Dumbarton–Fintry line may be related to deep-seated basement fractures.

The coalfield part of the district is traversed by numerous, generally east–west normal faults (Figure 16). An example is the Milngavie–Kilsyth Fault which bounds the coalfield to the north over much of its length. This fault trends slightly north of east and throws down to the south by up to 600 m. Most of the other larger faults, however, have somewhat different trends. The Priesthill and Barrhead faults both run roughly WSW–ENE and converge at Barrhead in a much shattered belt that continues to the WSW to form the Dusk Water Fault in north Ayrshire. The Castlemilk West Fault, which bounds the lavas to the west in the south-east corner of the district, trends NE–SW with a throw of some 400 m to the north-west. The Paisley Ruck has a NE–SW Caledonoid trend. It takes the form of a shatter belt 150 to 200 m wide, within which the strata are highly inclined (Hinxman et al., 1920, p.45). The resultant throw is down to the north-west but very variable in amount, reaching up to 550 m at Linwood. To the south-west it passes into several divergent faults, one of which continues its line into north Ayrshire. The continuation of the Paisley Ruck to the north-east as a fold is discussed above. The most extensive of the WNW–ESE faults is the Blythswood Fault but the largest throw is that down to the north-east along the Dechmont Fault, which locally separates Upper Coal Measures from Limestone Coal Formation. The amount known about the faults in the coalfield is distinctly variable and generally less than it is in the Airdrie district to the east. Mostly, their existence, general trends and approximate throws are known but, for example, the lines of the Balmore, Summerston and adjacent faults are rather speculative as are most of those in the little-mined zone on both sides of the River Clyde. The fault pattern in the Paisley area is also poorly known.

NINE

Quaternary

In the 19th century, the Quaternary deposits of the Glasgow district were described by amateur and professional scientists. Laskey (1822) described finding marine shells in clays excavated in the construction of the Ardrossan Canal. James Smith (1836; 1838; 1862) contributed seminal work on the coastal areas of Britain including the Clyde. During the Geological Survey's first survey in the 1860s, Bennie, Croll, Hull and Jack described the few permanent exposures and many temporary sections, and collected fossils and borehole records. Brady et al. (1874) independently described some of these excavations and exposures and the marine faunas they contained.

Papers by A Geikie (1863) on the glacial drift of Scotland and J Geikie (1872) on changes in climate during the glacial epoch were partly based on their experiences of the drift in central Scotland. Bennie (1866) was able to say that much of our knowledge of boulder clay was obtained from excavations in the hillock (drumlin) in the Bell's Park, Glasgow [5925 6575]. Although by this time, till (boulder clay) was correctly believed to be a glacial deposit, Wright (1896) and Neilson (1896) suggested a marine origin because of the microfossils they found in some excavations in Glasgow.

There was a strong tradition among members of the Geological Society of Glasgow of recording the geology of temporary excavations, ending with Wallace (1902; 1905). Examples include the cross striae described by Bell (1874) and Craig (1873), the presence of the bivalve *Mytilus* at 19 m OD on Oakshaw Hill, Paisley [4797 6400] (Fraser, 1873), and shell beds at Arkleston (Craig, 1877) and Dalmuir (Thomson, 1835). The more recent work is described in the review by Sutherland (1984), in Jardine (1986) and in Browne and McMillan (1989a).

The Scottish landmass has been glaciated on several occasions during the Quaternary Period with evidence for at least sixteen major cold events during the last 1.6 million years (Bowen, 1978; Price, 1983). Because of the erosive action of successive ice sheets only the events since about 30 000 radiocarbon years Before Present (BP) are known with some certainty and in any detail for the Glasgow district. Even during this period the record is incomplete, but from an assessment of evidence the glacial and postglacial history of the district can be constructed.

SUMMARY OF LATE QUATERNARY HISTORY

According to Price (1983) the most recent cold events took place from about 28 000 to 14 000 BP and 11 000 to 10 000 BP. Both events were associated with the build-up of glaciers in the mountains of the western Highlands of Scotland. The combination of increasing precipitation and cooling climate resulted in the generation of the Dimlington Stadial Ice Sheet which, at its maximum some 18 000 years ago, extended over most of the Scottish landmass as well as much of England. It is estimated that at this time the ice sheet was over 1 km thick in central Scotland. Till deposited from the ice covers a wide area, and locally conceals deposits of clay, sand and gravel and also earlier accumulations of till which were laid down some time prior to this stadial. The Dimlington Stadial Ice Sheet receded slowly over a period of several thousand years, due initially to a decrease in snowfall rather than a warming climate. That arctic conditions still prevailed in newly deglaciated areas between 15 000 and 13 500 years ago is indicated by the arctic fauna of the glaciomarine Errol Beds which were deposited in the valleys of eastern Scotland and offshore (Peacock 1975; 1981).

The final stages of recession of the Dimlington Stadial Ice Sheet were probably accompanied by the climatic amelioration which heralded the Windermere Interstadial (13 500 to 11 000 BP). By about 13 500 BP the Glasgow district was probably largely ice free. During deglaciation, rivers nourished by meltwaters transported and deposited sand and gravel, either under or adjacent to bodies of melting ice. Associated terraced landforms and kame and kettle topography developed at this time, especially in parts of the Clyde and Kelvin valleys. In the latter valley, meltwaters initially drained eastwards via the valleys of the Bonny Water and River Carron into the Forth estuary. In the former valley, meltwaters were initially ponded by north-westward retreating glacier ice occupying the Glasgow area. Deltaic sands and lacustrine clays and silts were deposited within the valley as far south as Lanark [880 440]. With further downwasting and recession of the ice, the barrier in the Clyde valley collapsed and the sea flooded the lower ground of the Glasgow area up to an elevation of about 45 m above OD. Initially marine clays of the Paisley and then the Linwood formations (the Clyde Beds of Peacock, 1981) were deposited and are preserved in parts of the Clyde, Blane and Endrick valleys. In the Glasgow district these sediments are fossiliferous (Plate 9) and, as elsewhere in the Firth of Clyde and Forth estuary (Browne et al., 1984), contain a fauna indicative of the cold temperate (Boreal) conditions pertaining to the Windermere Interstadial. By 11 000 BP local sea level in the Clyde valley had probably fallen from the late-Devensian maximum to present OD or lower.

With the return of arctic climatic conditions during the Loch Lomond Stadial between 11 000 and 10 000 BP, valley glaciers again developed in the western Highlands. The Loch Lomond glacier reached the north-western part of the Glasgow district damming in a glacial lake in the Endrick and Blane valleys. Glaciomarine sediments were deposited in the Clyde estuary and the Leven valley (Sheet 30W). There is evidence of periglacial activity (gelifluction) particularly in the Campsie Fells and on the flanks of drumlin slopes. Landslips in the Campsie Fells and the Kilpatrick Hills probably also developed during this period.

The climate started to improve about 10 000 BP and following initial fluctuations in relative sea level, as illustrated by the buried beaches of the Forth valley (Sissons, 1966; 1969), the main Flandrian marine transgression commenced about 8000 years ago. The maximum of the transgression in the Forth valley is placed at about 6500 BP (cf. Sissons and Brooks, 1971). Thereafter sea levels fell by stages to present OD. These changes in relative sea level may be seen in the Glasgow district where the upper limits of a series of terrace surfaces are recorded at 12 m above OD, at less than 7.5 m above OD and below 5 m above OD (Browne and McMillan, 1989a, fig. 5). Estuarine sands and gravels with clays and silts, associated with these levels, are present in Glasgow.

Most of the deposits flooring the floodplains of the principal valleys, including those of the Clyde and Kelvin rivers and Blane and Endrick waters, are of Recent age. During the Flandrian, basin peat and lake clays developed in interdrumlin areas and in hollows on higher ground. Hill peat also developed in the Campsie Fells, and the Kilpatrick and Renfrew Hills. Raised mosses were extensively developed on low ground north of Paisley. Locally, Flandrian peat may bury similar pre-existing late-Devensian deposits.

The following account of the Quaternary utilises the stratigraphical framework established by Browne and McMillan (1989a; 1989b) for the Clyde valley area with the addition of one newly defined unit, the Kelvin Formation (Forsyth et al., 1996). The formations are defined on the basis of sediment lithology and are classified according to their principal provenance. Glacial, lacustrine, fluvial, marine and organic origins are identified (Table 7). Principal localities are shown on Figure 17. The geographical distribution of the formations and their vertical relations are shown on the margin of the Glasgow Drift Sheet (BGS, 1994).

DEVENSIAN TOPOGRAPHY

On the high ground of the Glasgow district, where drift thicknesses generally do not exceed a few metres, the Devensian topography was probably similar to that of

Figure 17 Location plan for the Quaternary of the Glasgow district and adjacent areas.

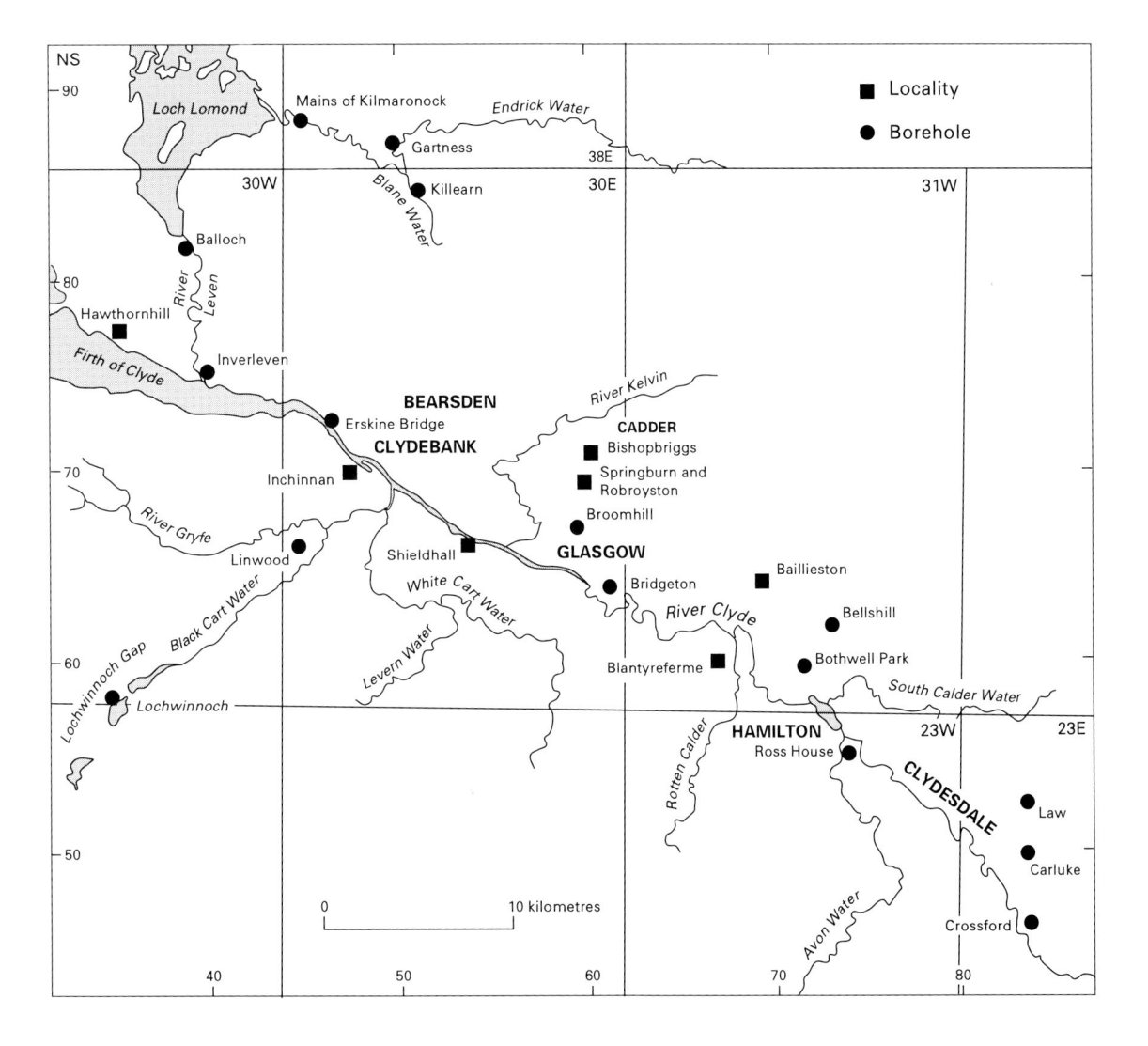

56 NINE QUATERNARY

a

b

c

d

e

f

g

h

i

j

k

l

Plate 9 Selected Devensian fossils.

a. *Buccinum undatum* (GSE 15098) × 1.0. Lands of Rylees Farm, Ralston, Paisley [513 645].
b. *Chlamys islandica* (GSE 15092 × 0.7. Langbank, east of railway station [331 785].
c. *Boreotrophon clathratus* (GSE 15095) ×2.0. Escavation at Gallowhill, Paisley [491 647].
d. *Nuculana pernula* (GSE 15091) × 2.0. Westmarch and Shortroods Clayfield, Paisley [4700 6500].
e. *Tridonta elliptica* (GSE 15089) × 1.0. Fullwood Brickfield, Houston [440 670].
f. *Hiatella arctica* (GSE 15090) × 1.5. Fullwood Brickfield, Houston [440 670].
g. *Mytilus edulis* (GSE 15097) × 1.0. Lands of Rylees Farm, Ralston, Paisley [513 645].
h. *Balanus balanus* (GSE 15094) × 2.0. Ralston Hill [509 640].
i. *Tectonatica clausa* (GSE 15093) × 1.5. Ralston Hill [509 640].
j. *Tridonta montagui* (GSE 15088) × 2.0. Fullwood Brickfield, Houston [440 670].
k. *Mya truncata* (GSE 15096) immature shell × 1.5. Lands of Rylees Farm, Ralston, Paisley [513 645].
l. *Macoma calcarea* (GSE 15099) × 2.0. Inchinnan Sewer Trench–west side [4761 6931].

today. The effect of the last ice sheet was to modify mainly pre-existing glaciated landforms and alluvial tracts, and redistribute till deposits. The erosive power of the ice is marked by glacial striae (Figure 18) and by rare *roches moutonées* and easterly aligned crag-and-tail landforms that are developed on resistant bedrock, particularly Lower Carboniferous igneous rocks. North of the Kelvin valley in the Campsie Fells, and also in the Kilpatrick Hills, till is thickest on the valley floors, with exposed rock and upland blanket peat predominating on the flanks and summits of hills. The Corrie of Balglass [590 850] and the Little Corrie [575 850] on the northern scarp of the Campsie Fells (Gregory, 1914) are thought to have been cut during the build-up of the Dimlington Stadial Ice Sheet (Figure 18).

On the low ground over much of the district, predominantly easterly moving Dimlington Stadial ice sculpted well-formed drumlins (Elder et al., 1935; Menzies, 1976; Rose, 1981), with generally west–east aligned axes (Figure 18). However, there is little doubt that major rockhead depressions under the valleys of the Clyde and Kelvin represent a generally pre-Dimlington Stadial topography which may have been initiated by river action in the Pliocene (cf. George, 1974, fig. 2.1) or earlier (cf. Linton, 1951). Bennie (1868) noted that the first indication of a deep channel below the Kelvin was an in-rush of sand into mineworkings for the Hurlet Limestone and Coal in the Braidfield Pit. The in-rush was at a depth of about 92 m and the miners escaped with the greatest difficulty. Bennie's field notebook records that the ground caved in to form a 'sit' on North Kilbowie Farm [5014 7144]. The surface level is about 49 m above OD. Bennie's notes also describe an in-rush of sand and water in Cullen's Pit near Bearsden [5314 7204]. This event happened during boring in sandstone when the rods unexpectedly dropped right down. Bennie concluded that the metals overhung the sand like a crag. The surface level of the site is about 77 m above OD and the depth about 102 m.

In the Airdrie district (Sheet 31W), the thickness of the sediments in the Kelvin depression range from about 30 m west of Kilsyth [700 770] to 87 m at Torrance [620 735]. Corresponding topographical levels for the bedrock surface range from 11 m above to 49 m below OD, illustrating a generally westwardly deepening depression. This trend is confirmed by rockhead levels of 50 to 75 m below OD recorded farther west in the district notably in the Bearsden [540 720] and Drumry [500 710] areas. This trend could indicate that erosion was initiated during a period of relative low sea level. However, such levels are lower than the mininum rockhead values recorded in the Firth of Clyde in the channel between Toward Point [135 670] and Skelmorlie [195 670]. Here Deegan et al. (1973, fig. 9) recorded levels of 40 to 60 m below OD. Such evidence points to overdeepening of the Kelvin bedrock depression by glacier ice or subglacial meltwaters under high hydraulic head.

In the Clyde valley east of Bothwell (Sheets 23W, 31W), rockhead level is as low as 6 m below OD and rises upstream. This shallow bedrock depression is probably separate from that which descends below OD downstream of Cambuslang [640 600] to Bridgeton [610 630], where boreholes show rockhead levels lower than 40 m below OD. Here, most of the infilling sediments are generally younger than the till of the Dimlington Stadial Ice Sheet. Information about the bedrock profile under the Clyde west of Renfrew, and under the Gryfe, the Black and White Cart waters, is too scarce to prove more than the general shape of pre-Dimlington Stadial depression under the district. Indeed there is insufficient information to show continuity of the bedrock depressions of the Clyde under Glasgow and the Linwood-Paisley embayment, with those of the Leven–Clyde confluence and the Kelvin. It is also likely that the deep depression under the Blane and Endrick valleys at Killearn could be separate from that under Loch Lomond.

Carruthers (*in* Clough et al., 1911) noted the coincidence of both the Kelvin and Clyde bedrock depressions with broadly synclinal structural axes in the underlying Carboniferous strata. However, he indicated that drainage may have been initiated on younger rocks which were subsequently eroded. Recent opinions on the origin of the depressions include those of Jardine (1977, pp.104–106) who proposed river action at a time of lower global sea level, and of Menzies (1981) who favoured excavation by subglacial streams. Jardine (1986, p.32) concluded that the rockhead depressions were probably composite both in origin and age. It seems most likely that erosive processes at work beneath ice sheets during the Quaternary have overdeepened these drainage systems.

PRE-DIMLINGTON STADIAL DEPOSITS

Bennie (1868) described the records of boreholes which proved over 100 m of drift in the Kelvin valley. He also reported a journal (unlocated) with five beds of till interbedded with sand and gravel and/or clay and silt.

Table 7 Lithostratigraphy of the Quaternary formations.

Main lithology	Clay and silt		Sand and gravel		Diamicton	Peat	Approximate age in radiocarbon years before present	
Provenance	Marine	Lacustrine and fluvial	Marine	Lacustrine and fluvial	Glacial	Organic		
FORMATIONS	Erskine	Kelvin	Gourock	Law		Clippens (earlier deposits exist locally)		FLANDRIAN STAGE
		Kilmaronock	Longhaugh	Endrick				
	Buchanan						10 000 (11 500)	
	Balloch	Blane Water	Inverleven	Drumbeg	Gartocharn Till		Loch Lomond Stadial	DEVENSIAN STAGE
							11 000 (12 800)	
	Linwood		Killearn				Windermere Interstadial	
	Paisley	Bellshill	Bridgeton	Ross				
				Broomhouse			13 500 (14 680)	
					Wilderness Till		Dimlington Stadial	
							27 500	
		Broomhill		Cadder	Baillieston Till (earlier deposits exist locally)		Pre-Dimlington Stadial	

Note: age in brackets is in calender years

Records in the BGS archives show three units of till separated by two of sand and gravel in boreholes at Garscadden [5310 7233] and Banner Road [5364 7028]. At least four tills are present in the Milngavie No. 4 [5599 7213] and the Boclair No. 6 [5688 7214] boreholes. Rockhead is at a depth of 105.3 m and 61.06 m respectively. The deepest record of drift (107.9 m) is in the Milngavie No. 5 Borehole [5735 7285]. Of the above boreholes, this is the only one which is sited on the valley floor (of the Allander Water). The others are situated on the flanks of drumlins or on valley sides.

The lithostratigraphy of the above boreholes is uncertain as is that of at least 30 other boreholes proving complex successions within the bedrock depressions of the Kelvin and Clyde valleys. Only two records mention the presence of shells. The Blairdardie No. 4 Pit [5246 7031] proved three tills separated by an upper thin deposit of shelly sand and a lower unit of mud and running sand, with rockhead at 51.82 m. Similarly the Drumchapel Pit [5223 7015] proved three tills. Between the lower two tills, was 10 m of laminated clay and silt with shells throughout, rockhead being recorded at a depth of 49.4 m.

Figure 18 Ice-flow and other glacial features in the Glasgow district.

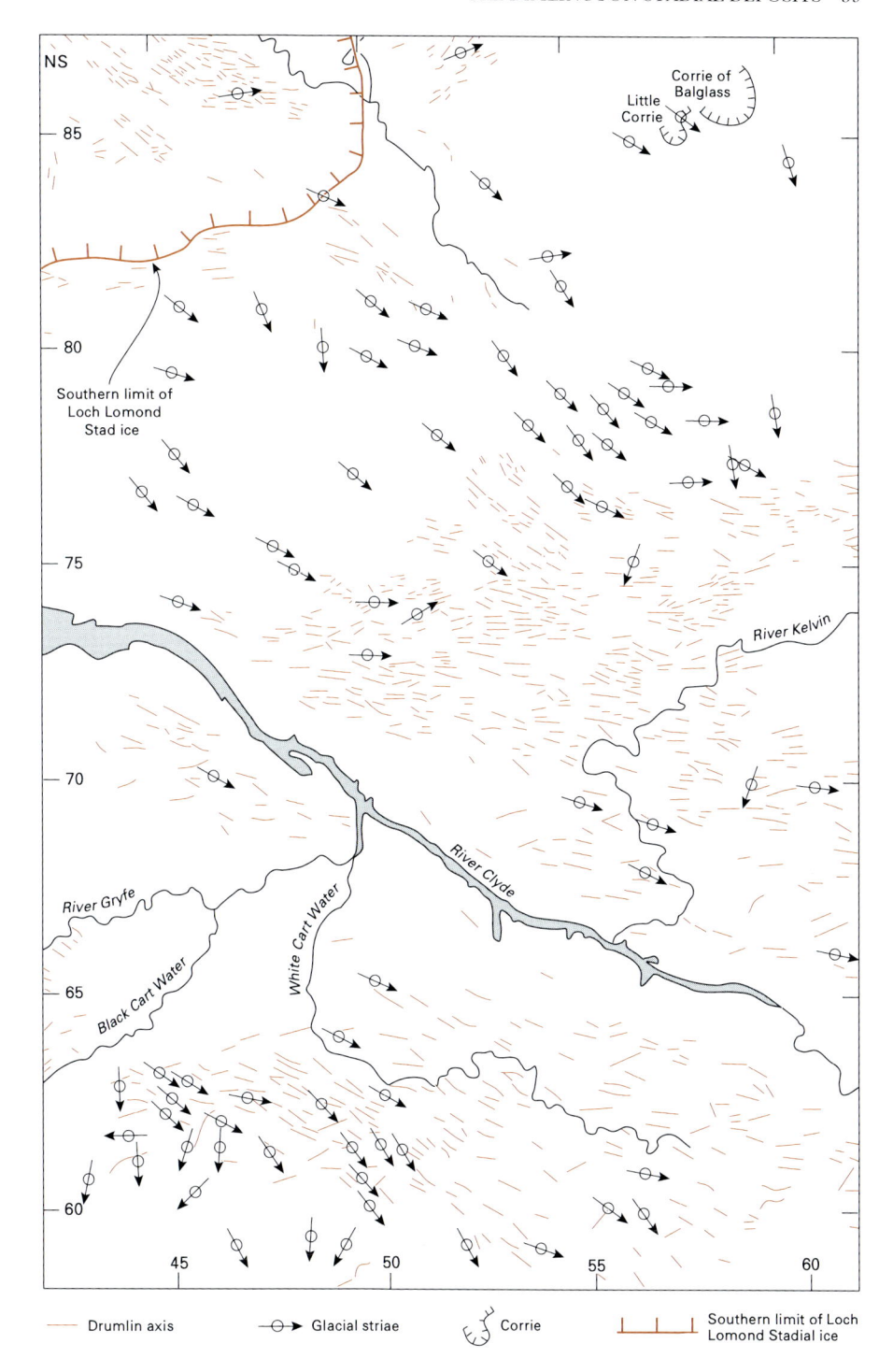

The oldest, pre-Dimlington Stadial sediments (BGS, 1994, section E) in the Glasgow district are preserved in a limited area around Erskine Bridge [465 726]. The lowest beds are a poorly known variable sequence of stiff mud (up to 4 m) and/or sand and gravel (up to 10 m) considered to have been deposited in an ice-dammed lake. They are underlain by glacial till (up to 3 m) resting on bedrock. These deposits are overlain by the relatively widespread Baillieston Till Formation. This till lies stratigraphically below the Wilderness Till Formation and, despite the absence of definitive age data, is considered to be the product of a glacial event earlier than the late-Devensian. The Baillieston Till is a glacially organised diamicton composed of boulders, pebbles and gravel in a matrix of sandy silty clay. Elongate clasts are preferentially oriented and the deposit is well jointed. In general the colour of the clay matrix reflects that of the local bedrock. Thus it is reddish brown at Erskine Bridge because of the presence of the nearby Upper Devonian bedrock. At Broomhill [598 665] it is greyish brown reflecting derivation from the glacial erosion of coal-bearing Carboniferous rocks.

Glaciofluvial outwash deposits

Outwash deposits related to the advance of the Dimlington Stadial Ice Sheet occur on the south side of the Kelvin valley in the Cadder area (BGS, 1994, section C). Here sand and gravel of the Cadder Formation rest on the Baillieston Till Formation and are in turn overlain by the Wilderness Till Formation. The poorly known subsurface sands and gravels and interbedded diamicton beds under the floodplain of the River Kelvin are also assigned to the Cadder Formation. Within the area of the Kelvin's bedrock depression the formation may be at least 44 m thick as illustrated by the Cadder No. 16 Borehole [6002 7326].

In the Cadder area, although exposures were formerly plentiful in workings for sand and gravel, there are now only two sections remaining where the Cadder Formation can be seen. The typical lithological assemblages consist of mainly framework-supported bouldery gravel and sand, and coarse- to fine-grained, sometimes pebbly, sand with silt. Locally a thin clay and silt bed of possible lacustrine origin (?Broomhill Formation) rests between the sand and gravel and the overlying Wilderness Till Formation. Based on observations from sections in the top 30 m of the formation, much of the gravel is thickly bedded and displays trough cross-bedding in sets up to 3 m thick. These interfinger with sands which are trough

Figure 19 Sections of the Cadder and Wilderness Till formations near Bishopbriggs.

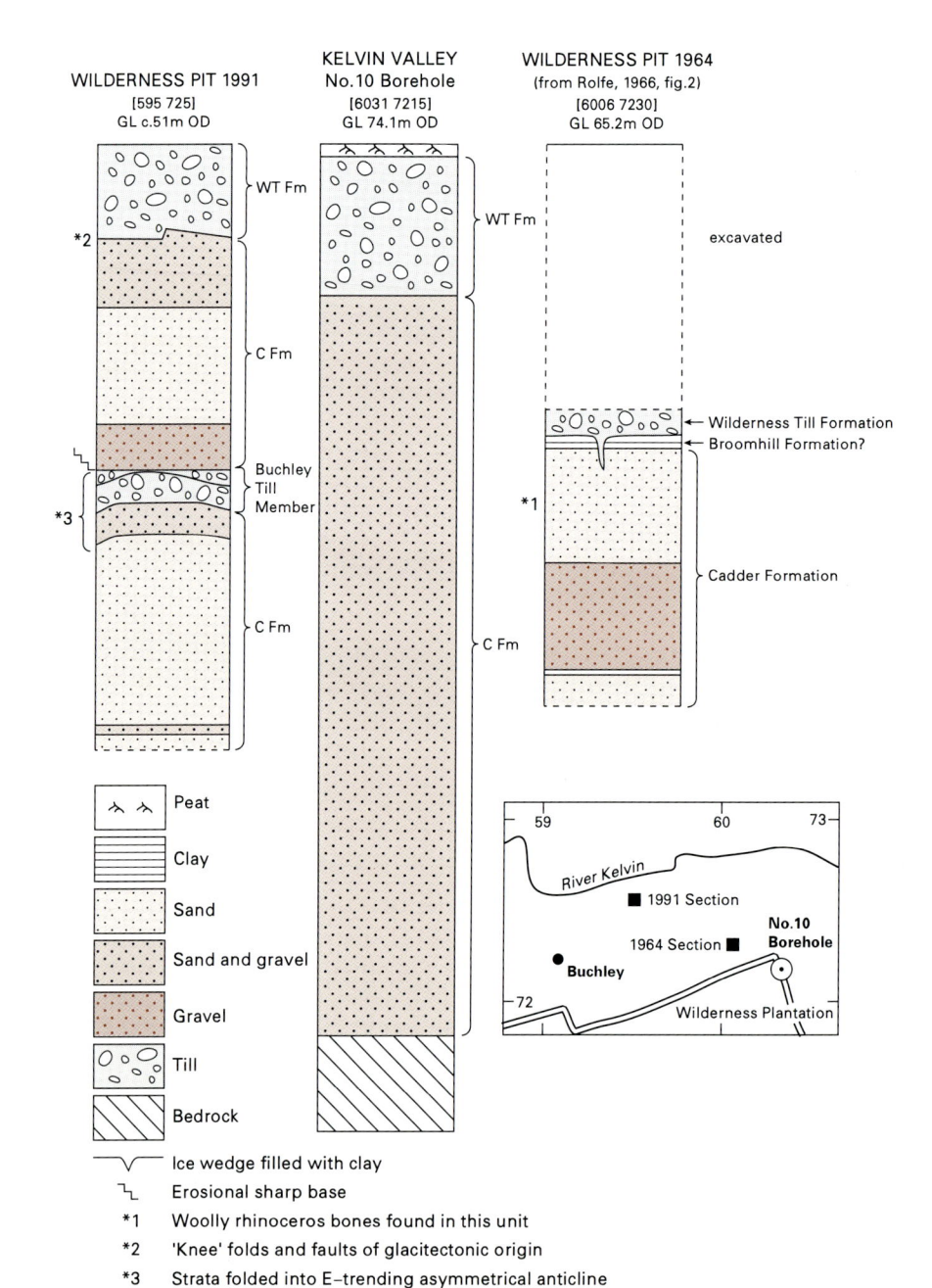

*1 Woolly rhinoceros bones found in this unit
*2 'Knee' folds and faults of glacitectonic origin
*3 Strata folded into E-trending asymmetrical anticline

cross-bedded, ripple laminated and horizontally laminated.

A recent temporary section in the Wilderness Gravel Pit, Bishopbriggs [595 725], is illustrated in Figure 19, together with an earlier record by Rolfe (1966, fig. 2) and the record of a recent borehole. The base of the Wilderness Till Formation is complex with shearing and incorporation of the underlying gravel. Locally there is a merging basal contact, sometimes with a basal deposit consisting of a matrix-supported sandy pebble gravel. Elsewhere, planar or even stepped basal contacts are developed. This diamicton is a deformation and lodgement till.

Faulting, folding and shearing of the Cadder Formation sediments below may be due in part to glacial overriding by the ice sheet that deposited the overlying Wilderness Till Formation. The faulting may also partly reflect the presence of buried masses of ice which have subsequently melted. Clay-filled fissures penetrating the top of the formation in the Wilderness Pit were observed by Rolfe (1966). These features were said to be common in this area and were regarded as ice-wedge casts which formed under periglacial conditions.

The precise status of the reddish brown and brown diamictons assigned to the Buchley Till Member in the Wilderness Pit (Figure 19) is uncertain. The base of this till is sharp and undulating (10 cm relief) and its top is erosive and planar. Possibly it formed during the ice advance which was responsible for the deposition of both the Cadder and Wilderness Till formations. The diamicton is thought to be glacial in origin and marks a possible local advance–retreat cycle within the earliest late Devensian. It may be much older (Baillieston Till Formation?). The underlying Cadder Formation sediments are folded, one 10 cm S-fold exhibiting a sheared core. All the beds between 6.50 and 8.40 m depth are folded into an east-trending anticline, the beds dipping to the north at 20–40°, and to the south at 10–20°.

From the Wilderness Pit [6006 7230], Rolfe (1966, pp.253–258) described the finding of bones of the woolly rhinoceros. One of the bones collected provided a radiocarbon date of 27 550 BP. This date is compatible with recently reported dates from a stratigraphically similar position from Sourlie [338 414], near Irvine (Jardine et al., 1988).

The Cadder Formation was formerly seen in a sandpit south-west of Bearsden Station [5390 7148], in a gravel pit east of Bearsden Station [5441 7183] and in a gravel pit at Ferguston Hill [5530 7223]. In each case the Wilderness Till overlies the granular deposits, being 3 to 4 m thick at the first site but only 1 m at the other two. Between 9 and 15 m of mainly large-scale trough cross-bedded sand with highly contorted clay bands were to be seen in the pit south-west of Bearsden Station. There the base of the overlying till consists of cemented gravel and is rich in boulders. Only 1 or 2 m of cross-bedded sand and gravel were seen at the other two localities.

The sedimentological and faunal evidence both point to deposition of the Cadder Formation in a periglacial environment with the possibility that some of the sediments are ice-contact deposits. The sediments appear to be fluvideltaic in origin, based on the scale of the trough cross-bedded units. Overall the deposits appear to form a major outwash system laid down in front of the advancing Dimlington Stadial Ice Sheet about 27 000 years ago. The outwash deposits were subsequently overridden by the ice which deposited the overlying Wilderness Till Formation.

Glaciolacustrine deposits

Pre-Dimlington Stadial lacustrine deposits are found in various areas around Glasgow. Subsurface data and temporary sections reveal that laminated muds rest discontinuously upon the Baillieston Till. Accorded the name Broomhill Formation, these sediments were proved in the BGS Erskine Bridge Borehole (Figure 20), which is taken as the standard section for this formation (from 16.31 to 32.75 m depth). Its stratigraphical position between the two tills is clearly demonstrated here. The typical lithology is thinly bedded silty clay with wisps, laminae and bands of silt and sometimes sand. The deposit is reddish brown in colour with grey or buff silt and sand. Isolated clasts, up to 14 cm, are considered to be dropstones. Thin diamicton bands and patches occur locally, with some graded bedding (turbidity flows) and microfaulting. Some polished surfaces and steeply dipping microfaults may indicate glacial disturbance by overriding ice. The laminations are probably varves, but with multiple layers, and may collectively represent 600 to 1000 years of sedimentation.

In Broomhill Park No. 2 Borehole [5985 6650] the formation (12.3–21.0 m) consists of dark brownish grey clays with reddish brown laminae and bands. Dips of greater than 30° are recorded in places and wisps and bands of diamicton (laminated in part) are common. The varved couplets are estimated to represent 1000 to 1500 years of sedimentation. Polished surfaces, joints and faults, together with folded bedding, all suggest glacial overriding.

Two sections in the New City Road area of Glasgow [5799 6743 to 5809 6744; 5781 6747 to 5790 6753] were recorded by E Hull in 1869. His notes describe up to 4 m of 'upper' boulder clay on what he said was 'in all probability the ordinary brickclay' (that is the Paisley Formation). The notes indicate that the upper till was indistinguishable from the ordinary older boulder clay (that is the Wilderness Till Formation) and the brick clay elsewhere rests on it. In the first section crumpling of the lamination of the clay was recorded at the west-end, with 'boulder clay inextricably confused'. It seems likely these structures are glacitectonic in origin and that the clays are part of the Broomhill Formation locally preserved below the Wilderness Till (that is the 'upper boulder clay'). The lower boulder clay is therefore part of the Baillieston Till Formation.

The evidence from sections and boreholes in the Broomhill Formation points to the deposition of the laminated muds in the form of varves representing seasonally controlled sedimentation in a temporary lake occupying part of the Clyde valley. Lake shorelines are estimated to have been in excess of 70 m above OD. Isolated clasts, interpreted as dropstones, and rafts of diamicton testify to the proximity of ice and it appears likely that the lake developed at some stage during the build-

62 NINE QUATERNARY

Figure 20 Sections of the Quaternary deposits in the Erskine Bridge, Linwood and Bridgeton boreholes.

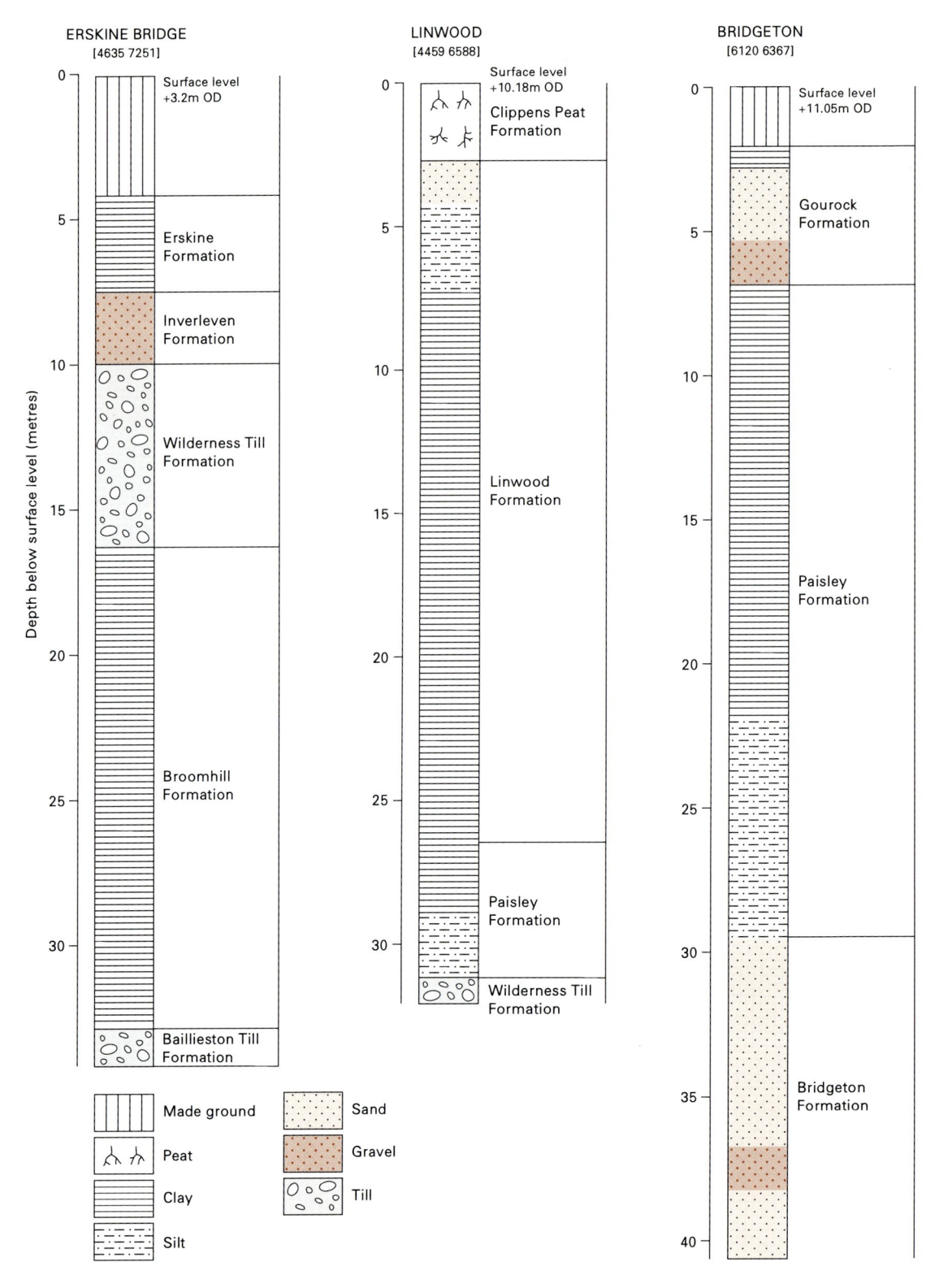

up of the Dimlington Stadial Ice Sheet. An estimate of between 600 and 1500 years has been made for the duration of lake sedimentation. The Broomhill Formation, as previously noted, may also be present in the Kelvin valley but the data are equivocal.

DIMLINGTON STADIAL ICE SHEET

Glacial deposits

Evidence from the Cadder area indicates that the build-up of the Dimlington Stadial Ice Sheet began at about 27 500 BP. At its maximum, around 18 000 BP, the ice sheet was probably over 1 km thick in central Scotland. The direction of ice movement is recorded by moulded landforms such as drumlins, some of which are rock-cored, and by crag-and-tail features, and by striae. Movement is generally from west to east with local variations from a north-westerly direction (Figure 18). In the Glasgow district the easterly movement of ice is confirmed by the distribution cone of erratic blocks of the distinctive Lennoxtown essexite (nepheline-monzogabbro) (Peach, 1909; Shakesby, 1976). The exposures of the essexite are located just to the east of Clachan of Campsie in the Airdrie district.

The principal deposit of the Dimlington Stadial Ice Sheet is the Wilderness Till Formation. Following the classification of Rose (1981) and Rose et al. (1988), the formation is named after the Wilderness Plantation north of Bishopbriggs [605 720]. In this area the formation comprises a hard, massive, reddish brown, sandy, silty diamicton which rests with stepped but low-angle disconformity on the cross-bedded sands of the Cadder Formation. At Drumry Wood [5093 7162] the contact was seen striking 350° and dipping 30°E. The diamicton contains clasts, the more elongate of which are preferentially oriented, and the deposit usually has systematic sets of joints. The colour of the matrix varies, depending upon the nature of the local bedrock. Minor features noted in the Wilderness Till include pockets and beds of medium-grained sand and thin beds of laminated clay. In the Cadder area, temporary sections in the Wilderness Till revealed that the diamicton is locally graded, displaying both upward and downward coarsening of the dispersed pebble- to boulder-sized clasts. Also recognised were zones within the formation in which the clasts become far less common. Locally the diamicton also appeared to have a 'structurally stratified' top with discontinuous partings sometimes of sand but elsewhere with no obvious lithological association. At Barhill Plantation [4620 7171] a horn of red deer was found in a bed of sand in stiff red till during the 19th century. Wright (1896) recorded the occurrence of marine calcareous microfaunas from the formation at a number of places, including Huntershill Quarry [607 695] and Hamilton Hill Brickworks [584 675].

Glaciotectonics

The effects of the Dimlington Stadial Ice Sheet overriding pre-existing glacial, fluvial and lacustrine sediments may be seen in sections at Cadder and Bearsden. In boreholes at Broomhill and Erskine Bridge cores exhibited steeply dipping faults and polished surfaces in geotechnically overconsolidated clay, silt and diamicton.

At Cadder, glaciotectonic disturbance in the Cadder Formation, which underlies the Wilderness Till Formation (Plate 10), is restricted to minor faulting and small-scale folding. In a former sandpit south of Bearsden Station [5390 7148], the undulating nature of the base of the Wilderness Till on the underlying Cadder Formation is thought to be partly glaciotectonic in origin. Neilson (1896) described 'seams' of coal (detritus?) in till in the Alexandria Parade area [616 654], just west of the district

Plate 10 Wilderness Till Formation on sand and gravel of the Cadder Formation, Wilderness Sandpit, near Bishopbriggs (D 126).

boundary. He noted that a seam was seen to 'rise at about an angle of 45° and turning across a sharp anticline descends at a similar angle'. He also recorded a 'seam' which curved, the upper end of which was nearly perpendicular. These structures may be glaciotectonic in origin, but also reflect the origin of the local diamicton as a deformation till.

WINDERMERE INTERSTADIAL: DEGLACIATION

Deposits of ice wastage

The initial stages of deglaciation were marked by ice wastage on the high ground of the Campsie Fells and the Kilpatrick and Renfrew hills. At first, summits of hills reappeared from beneath the ice. Continuing wastage in high valleys resulted in the deposition of 'morainic drift deposits', ill-sorted sediments comprising boulders and gravel in a sandy clay matrix. Hummocky morainic landforms are only developed in the Campsie Fells.

Glacial meltwater deposits

In time ice became restricted to the principal valleys and meltwaters issuing from glaciers transported and deposited large volumes of sediment. In west-central Scotland the direction of retreat was north-westwards towards the main Highland ice-source. As a result, in the Clyde valley ice receded progressively down-valley and seaward towards the inner Firth of Clyde and the fjords of the western Highlands. Likewise in the Kelvin valley, west of Kelvinhead [757 785] at the watershed between the westerly flowing River Kelvin and the easterly draining Bonny Water, the valley glacier receded westwards and down-valley. Nonmarine sand and gravel transported and deposited by meltwaters in both valleys are referred to the Broomhouse (glacial) and Ross (fluviodeltaic) formations.

Glaciofluvial deposits of the Clyde valley

The surface morphologies associated with the deposits of this formation are those of ice-contact. There are mounds and isolated flat-topped kames but no closed hollows (kettleholes). The lithological associations are rare matrix-supported bouldery gravel with sand, common framework-supported bouldery gravel with sand, and pebbly coarse- to fine-grained sand and silt. Overall, the most abundant deposit is sand, except where buried esker ridges are present. The deposits are up to 25 m thick. Much of the gravel is massive to crudely bedded but planar beds and trough cross-bedded units are also present. The sands are planar bedded, trough cross-bedded, ripple laminated and horizontally laminated. Deformed bedding, including reverse faults and folds, probably marks the former contacts between dead-ice masses and the sediments. Deformation is most common in the esker ridge deposits. The fine-grained sands and silts exhibit minor load casts and flame structures are present locally.

The Broomhouse Formation was formerly seen in excavations at the Erskine Bridge [4703 7286] where at least 2 m of glaciofluvial sand and gravel was overlain by 2.8 m of beach sand and gravel (Killearn Formation). This section and the associated ground ice (periglacial) structures were described and illustrated by Rose (1975, fig. 2). The formation was also seen in a sandpit adjacent to Clydebank Cemetery [4755 7285] by Carruthers in 1904. Beneath 1 m or so of red sand and gravel, now interpreted as a beach deposit (Killearn Formation), were several metres of slightly cross-bedded, red silty sand with thin clay bands. The whole succession was locally faulted and the bedding highly contorted.

A buried esker ridge (BGS, 1994, section B) was met with in the excavation of the Clyde Tunnel [542 662]. The esker is at least 15 m high and in cross-section 27 m wide at the top and 48 m at the base. The trend of the esker is approximately WNW with the southern and northern sides in contact with encasing sediments at dips of 57° and 45° respectively.

The sedimentological features of the Broomhouse Formation point to the operation of fluvial and deltaic processes. The coarsest gravel deposits are associated with esker ridges. They may have been deposited by glacial meltwaters flowing through tunnels in the wasting ice sheet. Massive deposits may have been laid down in the full-pipe sliding bed phase described by Saunderson (1977) but most show cross-bedding features consistent with fluviodeltaic processes and transport eastwards (data from the Airdrie district). The presence of sandy clay diamicton units may also point to the presence of decaying glacial ice but these might also have formed as debris flows in a fluvial context.

Lake Clydesdale: glaciolacustrine deposits

In the Clyde valley (Airdrie and Hamilton districts), upstream of Blantyreferme [673 595], a north-westward direction of deglaciation has been associated with the concept of an ice-dammed 'Lake Clydesdale' (Bell, 1871). Some authors, however, (e.g. Sissons, 1976, p.128) have preferred an up-valley, south-eastward direction of ice retreat in the Clyde valley. Browne and McMillan (1989a) and Forsyth et al. (1996) have identified proglacial lake deposits (Figure 21), the Bellshill Formation, the distribution of which appears to confirm the existence of such a lake (or series of lakes). The typical lithological association is of silty clay with laminae and beds of silt and sometimes sand. The deposit is dark brownish grey to brownish grey in colour. In the Glasgow district, the formation is known only from the Williamwood area [57 58] at an elevation of 46–53 m above OD. However, these deposits appear to have been laterally ice-wedged as they occur west of the position of the terminal moraine.

Glaciofluvial and glaciolacustrine deposits of the Kelvin valley

Flat-topped and moundy deposits of sand and gravel, assigned to the Ross and Broomhouse formations, occupy mainly the northern slopes of the Kelvin and Glazert

Figure 21 Deglaciation at the end of the Dimlington Stadial about 13 700–13 000 BP (after Browne and McMillan, 1989a).

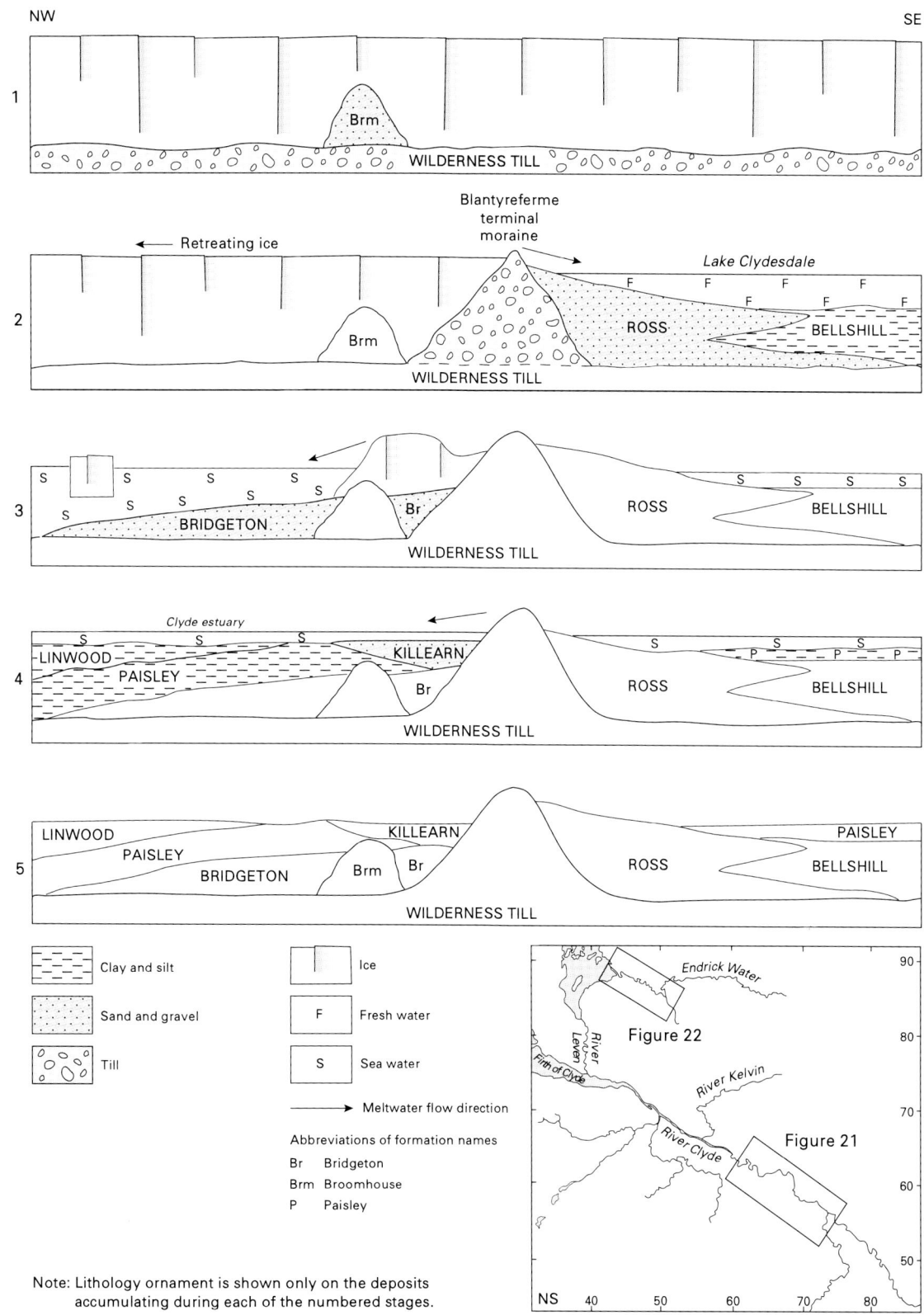

valleys and the south-east end of Strath Blane. Farther east in the Airdrie district, marginal drainage channels cut into some of these sands and gravels, and especially into the Wilderness Till and rock at topographically higher levels, indicating easterly drainage of meltwaters. These deposits form an integral part of the complex series of sediments which comprise the glaciofluvial and glaciomarine delta complex of the River Carron in the Forth valley (Browne et al., 1984).

South-west of Kilsyth (largely in the Airdrie district), a complex sequence of sediments flank the valley sides and infill the bedrock depression of the Kelvin. On the valley sides the sediments are disposed principally in the form of dissected terrace surfaces which are predominantly composed of sand with finely laminated silt beds and locally with gravel. The deposits which were once extensively worked at Torrance, Birdston and Cadder (Robertson and Haldane, 1937; Cameron et al., 1977) were considered to represent the margin of a glacial lake formed during a late phase of deglaciation as the valley glacier continued to retreat westwards (Clough et al., 1911). Recent borehole data tends to confirm the presence of both deltaic and lake bottom sediments which can be referred to the Ross and Bellshill formations respectively (BGS, 1994, section C). Locally ice-contact deposits of sand and gravel of the Broomhouse Formation are probably also present. These were formerly seen in a gravel pit at Langbank [5718 7313], where about 3 m of gravel and fine-grained sand displayed faulting and marked folding with subvertical downturns.

A succession through the glaciolacustrine sediments can be illustrated from Cadder No. 16 Borehole [6002 7326] where the Bellshill Formation is 17 m thick. It is overlain by 18 m of sand, part of which probably belongs to the Ross Formation. Little is known of the sedimentology hereabouts of the Bellshill Formation.

The presence of deltas building out into 'Lake Kelvin' is indicated by the sand and gravel bodies which have been assigned to the Ross Formation. Sediments deposited at the west end of the proglacial lake are interpreted as a complex of ice-contact and deltaic sands and gravels, assigned respectively to the Broomhouse and Ross formations. The transport direction of the Ross Formation was clearly displayed in temporary sections exposed at the Bishopbriggs No. 2 Sand Pit [625 734] which is just on the western edge of the Airdrie district. Here, large-scale trough cross-bedding indicated easterly current directions. Over 12 m of sand were formerly seen at this locality with beds of fine gravel at the top and silt layers near the base (?top of the underlying Bellshill Formation). Ice-contact association is indicated by the presence locally in these sediments of high-angle reverse and normal faults. The surface distribution of the Ross Formation at heights generally between 30 m and 55 m above OD is an indication of the lake level at the time this and the Bellshill Formation were deposited.

Ice clearance deposits: drainage of lakes Clydesdale and Kelvin

At about the time when Highland glacier ice in the Firth of Clyde had wasted sufficiently to allow 'Lake Clydesdale' to drain, the sea gained access to the Clyde estuary (Figure 22). As the ice front continued to retreat northwards as independent glaciers in the valleys of the south-western Highlands, any remaining ice in the Glasgow area and the Kelvin valley is considered to have become detached. Relative sea level in the inner Firth of Clyde was probably in excess of 35 m above OD at this time, some 13 500 BP (Browne et al., 1983b; Browne and McMillan, 1989a, pp.13–14). The Loch Lomond basin (Sheet 38W) and the 100 m-deep basin in Strath Blane (Doody et al., 1993) would also have been cleared of ice by northerly retreat. This would have happened whilst sea level was at about its highest, allowing the late-Devensian sea to gain access to both these areas not much later than in the Clyde and Kelvin valleys.

If observations and deductions about the altitudinal limits of marine (>35 m OD) and lacustrine (>70 m OD) deposits are reasonably accurate for the lower Clyde valley either side of Glasgow, failure of the ice dam at the Blantyreferme terminal moraine (Figure 22; Browne and McMillan, 1989a) would have released lakewaters under a considerable hydraulic head. These waters would have had great potential for erosion of pre-existing sediments and for deposition farther downstream. In particular, large volumes of sand and gravel would be available for erosion from the Ross and Broomhouse formations in eastern Glasgow. Redeposition of these is believed to have produced the Bridgeton Formation.

The Bridgeton Borehole [6120 6367] contains the standard but incomplete section for this formation between 29.31 m and the base of the borehole at 40.38 m (Figure 20). The two typical lithologies recognised are beds of very fine- to medium-grained (sometimes coarse-grained) sand, and fine to coarse gravel and boulders with a sandy matrix. Some of the sand was seen to be flat bedded but no other sedimentological features were identifiable. A nearby site investigation borehole recorded the base of the formation at 50.29 m, resting on the Wilderness Till. The formation is mainly unfossiliferous.

Sections in the formation at Shieldhall [536 664] have been recorded by McMillan and Browne (1989; Figure 23) who commented on evidence for former buried ice masses, and noted sedimentological features reminiscent of those described by Rust and Romanelli (1975) and Cheel and Rust (1982) for subaqueous outwash fans forming at an ice-front. A major difference between the Bridgeton Formation and the Broomhouse and Ross formations is that the deposits at Shieldhall were transported by westward flowing currents. This indicates that the associated ice upstream in the Glasgow area was probably detached from the active glacier ice which was retreating towards the western Highlands. Whether dead ice in eastern Glasgow could allow the transmission of such volumes of sediment and meltwater is open to debate but a source for the deposit is available in the sediments of the Broomhouse and Ross formations farther upstream. A source of meltwater could be the rapid draining of 'Lake Clydesdale' in which the Bellshill and Ross formations had been laid down. McMillan and Browne (1989) interpreted the Bridgeton Formation on

Figure 22 Glaciation and deglaciation of the Endrick valley area during the Loch Lomond Stadial about 10 800–10 000 BP (after Browne and McMillan, 1989a).

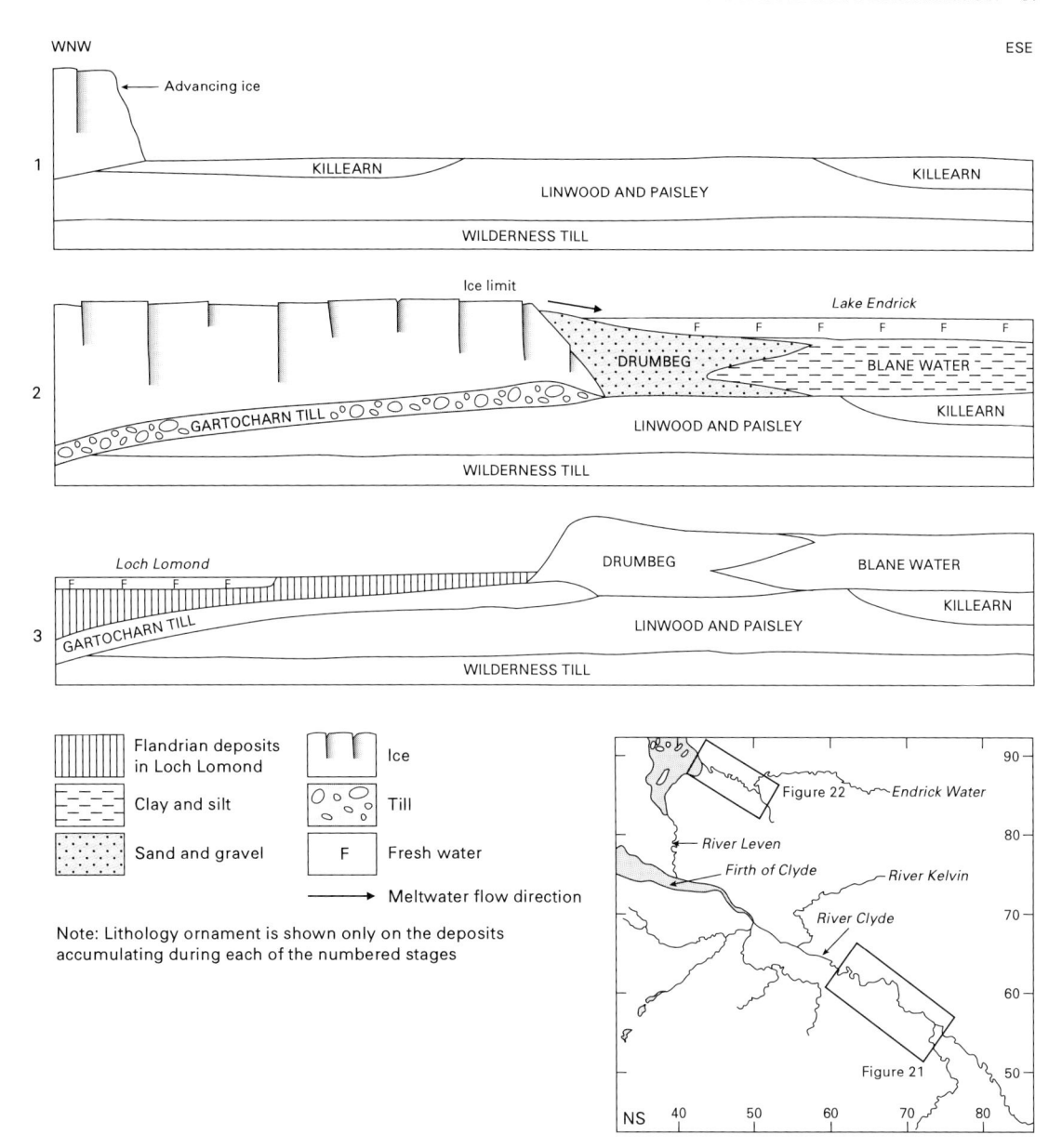

this basis, as an outwash deposit formed in a submarine environment. The slump and load structures described by Jardine (1965, figs 1 and 2) in fine-grained sand at St Vincent Street [5811 6567] in central Glasgow could be easily matched with similar ones seen at Shieldhall in the Bridgeton Formation. This suggests that Jardine's section was cut in Bridgeton and Paisley formation deposits.

Dead ice must also have remained to cap some of the remnant mounds of sand and gravel of the Broomhouse Formation. Locally the mounds would have been situated at elevations low enough to allow burial by the marine clay and silt of the Paisley Formation. As at Shieldhall, the Greenoakhill sections in the Airdrie district reveal evidence for burial of ice which survived perhaps for many hundreds of years in or under deposits of sand and gravel sealed under a thick cap of marine clay and silt of the Paisley Formation. The bedding of the overlying muds is disrupted by normal and reverse high-angle faults, and by low-angle thrusts with folding. These structures developed when the ice, buried in the underlying Broomhouse and Ross formations, melted.

Shieldhall and Greenoakhill are not the only localities where evidence for the survival of buried ice beneath marine sediments may exist. Bennie (1866) observed sections (Figure 24) at Windmillcroft Dock [583 647] where the base of a marine clay (Paisley Formation) always rests irregularly in large hollows on the underlying fine-grained, white sand (Bridgeton Formation). The sand contains rare isolated boulders some of which were polished and striated. The lamination in the marine clay was observed to be inclined at high angles and to be cut by sand dykes. Wallace (1905) at the Trongate [596 649] recorded dips of over 30° in finely laminated clay with sand partings (basal Paisley or top of Bridgeton Formation). Robertson and Crosskey (1874, p.246) reported a major excavation (Figure 25) in the Stobcross railway cutting [5639 6574 to 5718 6554] in which 'violent contortions' in the clay (Paisley Formation) were seen bent

68 NINE QUATERNARY

Figure 23 Horizontal section, Shieldhall, Glasgow. Till and rock were proved in boreholes, overlying deposits were seen in temporary sections.

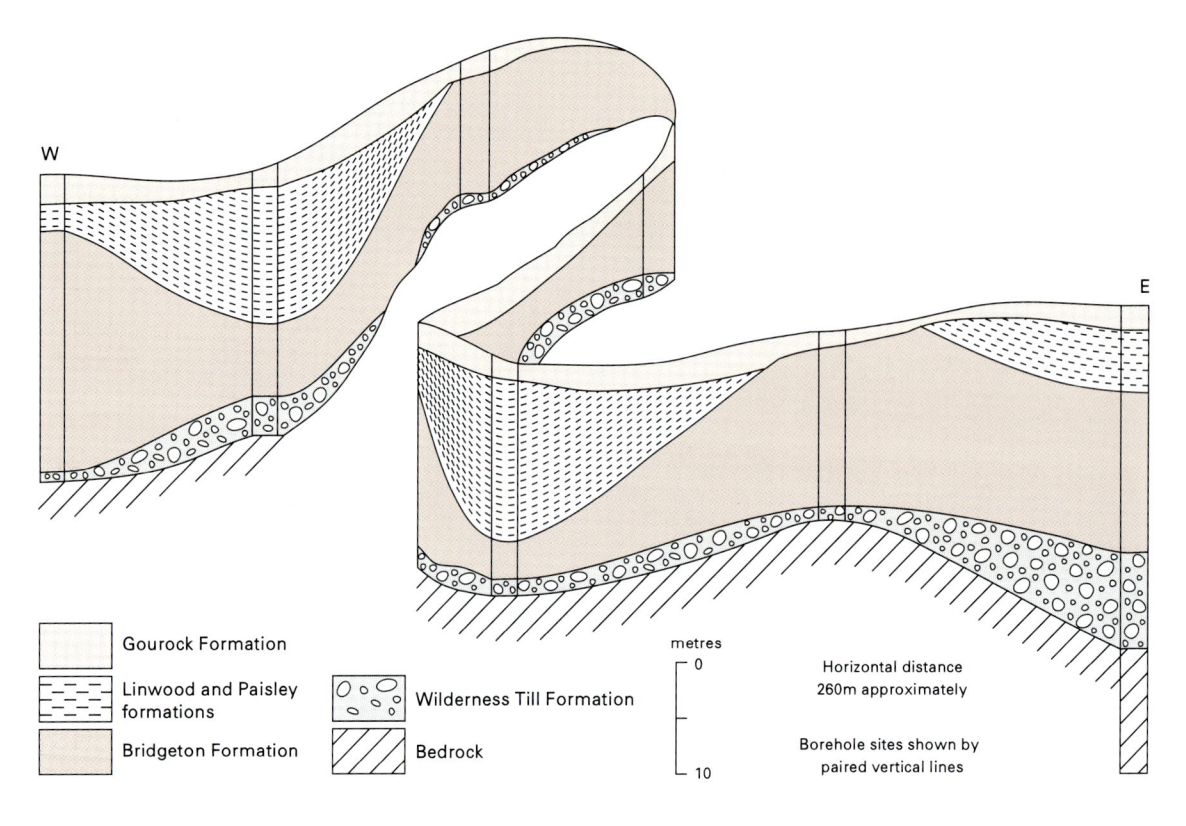

Gourock Formation
Linwood and Paisley formations
Bridgeton Formation
Wilderness Till Formation
Bedrock

Horizontal distance 260m approximately
Borehole sites shown by paired vertical lines

Figure 24 Generalised section in Windmillcroft Dock (after Bennie, 1866).

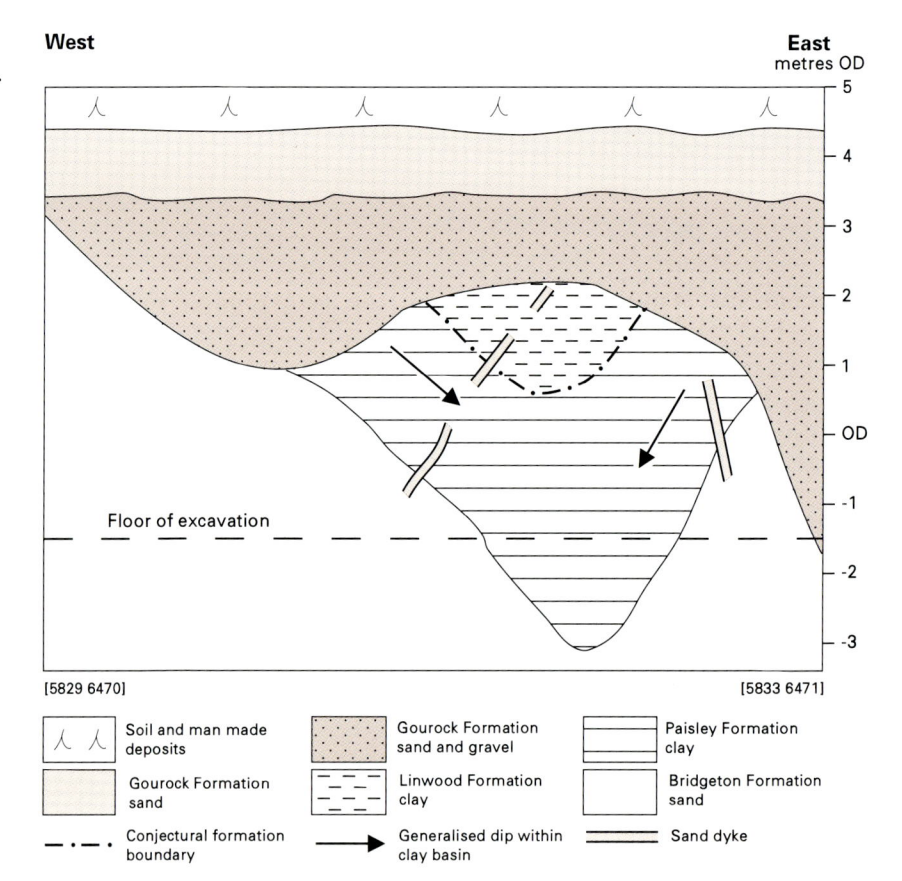

Soil and man made deposits
Gourock Formation sand
Conjectural formation boundary
Gourock Formation sand and gravel
Linwood Formation clay
Generalised dip within clay basin
Paisley Formation clay
Bridgeton Formation sand
Sand dyke

Figure 25 Section at Stobcross railway cutting, north side, not to scale (after Robertson and Crosske, 1874).

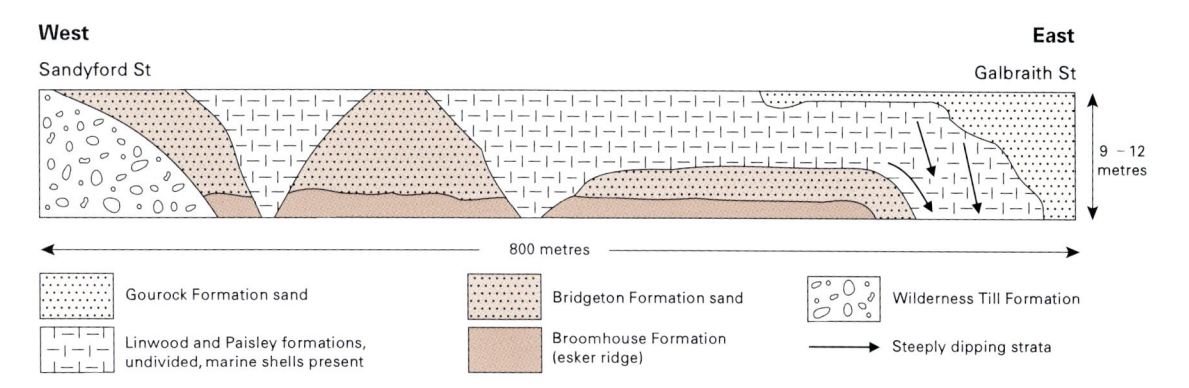

into deep hollows or troughs into the underlying sand (Bridgeton Formation).

Marine inundation

Peacock (1971) suggested that the route of entry of the late-Devensian sea to the Paisley–Glasgow basin was not via the Clyde estuary but by a more southerly route, through the Lochwinnoch Gap between Johnstone [43 63] and Dalry [29 49] (Sheet 22E). There is no stratigraphical evidence to support this suggestion despite two attempts to find the linking marine deposits by drilling: at North Kerse [339 556] Dickson et al. (1976) and at Lochwinnoch [352 581] (Browne and McMillan, 1989a, pp.44, 61). In the absence of either stratigraphical or geomorphological evidence to the contrary, it is thought that dead ice blocked the gap until after deglaciation of the Glasgow area.

Once the water levels of 'Lake Clydesdale', 'Lake Kelvin' and the sea had reached equilibrium in the Clyde valley area, high energy environments were restricted to the beach zone and to the mouths of tributary valleys where lateral deltas built out. Initially, meltwater plumes dominated and the Paisley Formation reflects this influence. The formation was proved in the BGS Bridgeton [6120 6387], Killearn [5100 8467] and Linwood [4459 6588] boreholes. The Linwood Borehole (Figure 20) contains the standard section where the formation (26.5–31.1 m) consists of four lithofacies. The uppermost consists of clay and silt (26.5–28.14 m), finely colour banded in shades of brownish grey, greyish brown and grey, that contain a sparse calcareous marine microfauna and macrofauna. The second is silt with clay bands (28.14–29.20 m), that is grey, well bedded, and contains a sparse calcareous marine microfauna. The third is clay with silt layers (29.20–30.23 m), very finely colour laminated in shades of reddish brown, dusky red, orange and grey. The lowermost consists of grey, well-bedded silt, with some sand and clay beds (30.23–31.10 m). This sequence is typical of the Clyde valley west of Govan and the Loch Lomond area.

The succession is a lot thicker to the east of Govan, and the formation in the Bridgeton Borehole (Figure 20) consists of three lithofacies. The uppermost is silty clay, with many silt laminae and thin beds (6.9–16.23 m), which is colour banded in shades of brownish grey and sometimes reddish brown. Overfolds, normal and low-angle reverse faults, primary current lineation and burrows are present. The middle is silty clay with many silt laminae and thin beds (16.23–21.77 m), which is colour banded in shades of brownish grey and reddish brown with beds of finely colour laminated clay in shades of reddish brown, brown and grey. The lowermost is silt, with clayey silt and clay bands (21.77–29.31 m), which is very thinly laminated to thinly bedded, brownish grey, reddish brown and grey. Slumped bedding including overfolds is developed. All three lithofacies are rarely fossiliferous.

Other sections in the thin (western) variant of the Paisley Formation include the temporary sections at Shieldhall. Here the lithofacies sequence resembles that of the Linwood Borehole (Browne and McMillan, 1989a, p.24). Sedimentological features recorded included pipe-like burrows, sweep and burst laminations (Mackiewicz et al., 1984), graded units, normal and reverse microfaults, small rip-up clasts (less than 1 cm) and diamicton wisps up to 5 cm long by 5 mm thick. In the Killearn Borehole [5100 8467] simple burrows, microfaults, primary current lineation and, as in most sections, dropstones were seen. In Mansionhouse Road, Paisley [492 642] a temporary section revealed 45 cm of Paisley Formation, the highly laminated clay and silt resting on the eroded top of the Wilderness Till Formation and dipping to the south at an angle of 16°. The upper surface of the laminated deposit is also eroded with isolated parts of laminae preserved at the contact with a ripple-laminated silt (15 cm thick) forming the base of the overlying shelly Linwood Formation (P Aspen, written communication).

Some of the sedimentological features of the Paisley Formation suggest deposition by traction currents (turbiditic density underflow); these include graded bedding, primary current lineation, erosion scours and rip-up clasts. A turbidite origin is more likely to explain the localised ponded distribution of these deposits commented on by McGown and Miller (1984) and interpreted by them as evidence for deposition in isolated lakes. Piper et al. (1983, fig. 11) illustrate acoustic facies in fjord basins where ponded distribution of sediments may be related to a turbidity-flow origin.

The most striking feature, however, is the fine-scale layering of some of the clays. This has been attributed to seasonal banding (varves) in proglacial lakes, a view most recently stated by McGown and Miller (1984), but Jardine's review (1986) of the general stratigraphy of the Clyde estuary states the now widely held belief that they are glaciomarine in origin. This view is compatible with the faunal evidence, even if the scarce marine assemblages of Glasgow and Paisley areas were derived by erosion from older sediments, for it is unlikely that the marine assemblages at the Hawthornhill site by Dumbarton [375 760] and at Ardyne [098 683] in the Firth of Clyde (Peacock et al., 1978) are anything other than contemporaneous with the sediments in which they are found. Such fine-scale layering has been reported from glaciomarine environments close to fjord glaciers. Mackiewicz et al. (1984) believed that the origin of the layering in their interlaminated facies was due to flocculation from overflow or interflow meltwater plumes interacting with a sea-water wedge that fluctuated in response to the tidal cycle. A similar origin for the Paisley Formation is possible, in effect an ice-clearance facies deposited in a salinity-stratified estuarine environment. Rapid sedimentation is likely because the evidence for bioturbation of sediments is slight, with only a few examples of trails and burrows recorded, except in the uppermost part of the succession in the Linwood Borehole and Shieldhall sections.

There is a major disparity in thickness between the Linwood and Bridgeton developments of the formation and, although the lithofacies associations are generally similar, they are present in different proportions. The difference is believed to be due to a much higher rate of sediment supply to the depositional basin east of central Glasgow.

The marine fauna is substantial in the more seaward parts of the Clyde region (Peacock et al., 1978; Browne et al., 1983b; Browne and McMillan, 1989a, pp.23–26) but sparse farther inland, in the Glasgow–Hamilton area, where sediment supply was greater. The maximum altitude to which clays assigned to the Paisley Formation are known to occur is generally over 40 m above OD in eastern Glasgow. Possible equivalent but unfossiliferous deposits at Dougalston Loch [561 738], to the north-west of the Kelvin valley, attain an elevation of 41 m above OD. However, sediments even tentatively identified as belonging to the Paisley Formation have yet to be located farther east in the Kelvin valley.

In the Glasgow and Paisley areas the Paisley Formation is commonly overlain by thickly bedded muds of the Linwood Formation (Browne and McMillan, 1989a, pp.26–27). This formation was proved by the Linwood and Killearn boreholes. The Linwood Formation is sulphide rich, and contains debris rafted by ice and some wisps of diamicton. Air-dried samples of the mud suggest the presence of vague and discontinuous silty or micaceous wisps which may have been formed in the manner described as 'sweep and burst texture' by Mackiewicz et al. (1984). Elsewhere there is evidence of strong reworking by burrowing.

The Linwood Borehole contains the standard section for this formation (2.64–26.5 m) and shows three lithofacies. The uppermost consists of silt and clay, medium grey and micaceous (2.64–6.65 m). Sand occurs at and near the top, in layers up to 3 cm thick. Seaweed and other plant remains are present, with eroded and stained foraminifera, and traces of rooty horizons near the top. The middle is of clayey silt and silty clay (6.65–18.85 m) which are medium to dark grey becoming brownish grey at depth. Silt and sand layers (and beds) and black sulphide beds and mottling are developed. Seaweed and marine shells are present. The lowermost is clayey silt and silty clay (18.85–26.5 m), brownish grey in colour but sometimes medium grey, thickly bedded to massive, with silt dustings and wisps or indistinct partings. Many marine shells are present and there is some sulphide mottling. The beds are heavily burrowed. It is possible that the middle unit of this borehole is the same deposit as the dark grey clay recorded by Robertson and Crosskey (1874) at Fairfield Dock [546 662]. In the Killearn Borehole (34.8–45.41 m) only the lowest unit of the Linwood Formation is present. The deposit contains rare beds of diamicton usually less than 5 cm but in one case up to 36 cm thick. Some sand layers and sharp-based beds up to 18 cm thick (?storm beds) are present.

The Linwood Formation was deposited as sea-bottom muds in moderate to shallow water depths in a well-mixed estuarine environment. The presence of isolated gravel to pebble-sized clasts interpreted as ice-rafted dropstones indicates the presence of shore ice in winter as it is believed the whole area was free of glacier ice at the time of deposition. Overall the succession coarsens upwards as the marine embayment became shallower and more estuarine in character. The presence of black sulphide beds and mottlings indicates anoxic sea-bottom conditions. This probably reflects the infilling of the marine embayment coupled with lowering of the relative sea level and the existence of a fjordic sill at the mouth of the basin around the narrows of the Erskine Bridge area. The heterolithic upper unit of the Linwood Borehole, with evidence of rooty horizons and eroded and stained foraminiferan tests, is probably intertidal in origin. The roots may be associated with the Clippens Peat Formation above.

At a number of localities, the Linwood Formation is sandy or contains beds of gravel rich in shells. An example of the former is the section on the M8 Motorway east of the Renfrew Road Bridge described by Aspen and Jardine (1968, fig. 1). Here at least 3 m of shelly silty sand (Bed 2) were seen at their localities A [4937 6571] and B [4967 6570]. Examples of gravel rich in shells were described by Browne et al. (1977) at Inchinnan (Figure 26). At one locality [4766 6940] the shelly gravel is fine to coarse (0.075 m maximum size), angular to subrounded, with a sandy and clayey matrix. A valve of *Arctica islandica* was dated at 13 142 BP whereas a valve of *Modiolus modiolus* gave a younger date of 12 011 BP. The ground level is 11.0 m above OD, the base of the gravel at 6.1 m above OD, and the top at 7.22 m above OD. The gravel at Inchinnan is overlain by fine to medium-grained, yellow brown sand, with thin brownish grey silty clay laminae, and rare isolated subangular stones up to 0.05 m. This deposit is flat bedded with a low-angle dip towards the centre of the local valley. Vegetable matter is present near the base. On top of the sand is a brownish

Figure 26 Generalised section of the Quaternary deposits at Inchinnan.

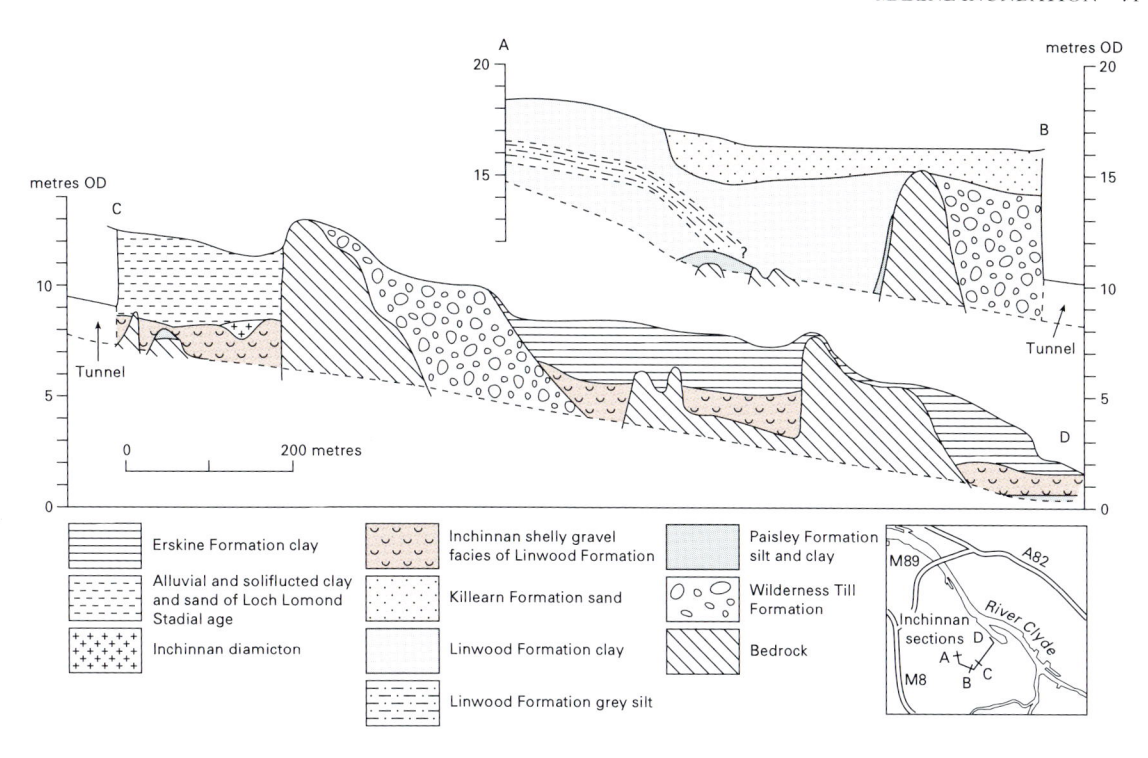

grey or greyish brown clay or clayey silt, with many thin bands of yellow-brown fine-grained sand. Flame textures and load casts are present at the tops and bases of sand beds. Isolated subangular stones up to 0.05 m are scattered in the clay. Both the clay and the underlying sand may be a mix of soliflucted and alluvially deposited sediments.

The beds of gravel or gravelly clay in the Inchinnan and other sections appear in marginal marine, sublittoral situations in which higher energy conditions prevented the deposition of fine deposits or gently reworked them (Peacock, 1989). The Inchinnan localities are all geographically close to conspicuous shoreline features of late-Devensian age at and below 15 m above OD which are less than 12 000 years old.

Numerous radiocarbon dates from in-situ and derived shell material originating in the most fossiliferous basal part of the Linwood Formation yield a spectrum of ages between 13 150 and 11 300 BP. Sedimentation was intermittent in the Linwood Formation and, as with the present sea floor, deposition of mud, sand or gravel, or growth of shell beds, was taking place in some places while elsewhere there was erosion or non-deposition.

There is some evidence for locally variable contacts and relationships between the Linwood, Paisley and Wilderness Till formations and bedrock. Anderson in 1938 observed sewer trench and tunnel sections at the Cart Harbour [4844 6526 to 4855 6556] and noted that over a short distance the Linwood Formation rests directly on both of the other formations and on bedrock.

The fauna of the Paisley and Linwood formations, Appendix 6, includes cold-water indicators such as *Chlamys islandica, Macoma calcarea, Nuculana permula* and *Elphidium clavatum*, reflecting a harsh but not fully arctic glaciomarine environment. The greater abundances and diversity of the fauna in the Linwood Formation suggests that conditions were a little more favourable and more fully marine than those prevailing during the deposition of the Paisley Formation.

The Killearn Borehole contains the standard section in the Killearn Formation (15.61–34.8 m). The chief lithology is fine- to medium-grained sand. Units of fine-grained sand with silt or clay layers also occur, especially near the top and base, and beds of fine gravel or sand with gravel (3 cm maximum clast size) are present in the upper part. The beds of the formation are reddish brown or orange in colour. The deposit forms an upward-coarsening deltaic unit from the base at 34.8 m up to 17.95 m and then fines up to the top of the formation at 15.61 m. Unfortunately the basal contact with the Linwood Formation was not seen due to core loss.

Similar deposits are present locally in the Clyde valley in central and eastern Glasgow, as proved in commercial boreholes, for example near the Bridgeton Borehole. They may include deposits of beaches and river terraces as well as deltas. The stratigraphical relationships of this formation are not well known, but are partly illustrated by sections at Scotstoun House [525 680] described by Jardine and Moisley (1967, fig. 1), by Rose (1975) at Erskine Bridge [4703 7286] and also in the temporary section by Clydebank Cemetery [4755 7285] seen by Carruthers in 1904. In the last two sections the beach deposits are well-bedded sand and gravel.

The sea-level curve for the Clyde area following deglaciation is poorly known compared with that for the eastern coast firths (Sissons, 1976, fig. 9d; Armstrong et al., 1985, figs 13, 15). In the Rutherglen [610 618] and Carmyle [640 620] areas deltaic sand and gravel deposits of the Killearn Formation occur. Elsewhere in Glasgow, the Firth of Clyde and Strath Blane, raised shorelines, raised beach and delta deposits are present but no clear assessment of relative sea-level changes can be reconstructed in the absence of systematic levelling. It is likely, as in the Forth valley, that, during the Windermere Interstadial and by the time of the Loch Lomond Stadial, the relative sea level fell to OD or below, due to the local dominance of isostatic uplift of the land over the generally rising world sea level.

The only lacustrine deposits of equivalent age to the Linwood Formation which have been studied in any detail are to be found at Garscadden Mains [5327 7059]. Here Mitchell (1952) described 2.15 m of muds with sand and clay and proposed a stratigraphy for this classic site. Anderson and Simpson added an addendum on the marine origin of the varved clay now assigned to the Paisley Formation. Jardine (1969) described further sections near Garscadden but these did not encounter the lake muds.

On land, the contemporary pioneering vegetation appears to have been dominated by grasses, sedges and ferns. This heathland, mossy in part, contained patches of bare ground or disturbed soil as indicated by pollen of *Rumex* and *Plantago*. These two forms may indicate near periglacial conditions but limited snow cover.

LOCH LOMOND STADIAL

About 11 000 to 10 000 years ago, there was a short-lived glacial event in Scotland, the Loch Lomond Readvance (Simpson, 1933). From its source area in the western Highlands, glacier ice extended into the district to just south of Finnich Glen [49 84] in Strath Blane. Associated with the ice was deposition of the Gartocharn Till and the Drumbeg Formation sand and gravel. Within the district, the event, modelled in BGS, 1994, section A (Figure 22), is mainly marked by the deposition of marine muds (Balloch Formation) in the Vale of Leven and outer Clyde estuary (Sheets 29E and 30W). These contain the arctic bivalve *Portlandia arctica*, a shell commonly found in the muddy environment near a glacier snout. The event is also marked by the deposition of varved muds (Blane Water Formation) in an ice-dammed lake in Strath Blane and the Endrick valley. The maximum southern extent of the ice is generally shown by terminal moraine landforms, on higher ground usually composed of till (Gartocharn Till Formation) but on the lower ground of sand and gravel (Drumbeg Formation). Whether the terminal moraine features mark the maximum extent of the ice in all cases is open to question, following the discovery of blocks of shelly Linwood Formation clay (Browne et al., 1983a) in the Gartocharn Till at Townhead [3161 8289] east of Helensburgh (Sheet 30W). The locality is 1.1 km south of the terminal moraine and south of the watershed between Glen Fruin and the Clyde valley.

The Gartocharn Till Formation is seen in field sections in the vicinity of Gartocharn by Loch Lomond, and the name is based upon the classification of Rose et al. (1988) and Rose (1981; 1989). The Mains of Kilmaronock Borehole (Figure 27) contains the standard section (28.07–32.99 m). The lithology displayed in this borehole consists of massive silty clay with isolated pebbles, cobbles and boulders and knots of sand. This diamicton is brown or dark brownish grey with many dark sulphide patches and marine shells. It contains joints and low-angle thrust surfaces, some of which are polished, striated and grooved. Locally rafts and more diffuse patches of bedded clay are present. The general characteristics of this deposit are owed almost entirely to the nature of the clays of the Linwood Formation which have been glacially reworked by the ice to form the till.

Around Gartocharn many small sections confirm the existence of a somewhat more stony shelly till than that seen in the Mains of Kilmaronock Borehole. Jack (1875) described sections of shelly till in the Catter Burn [4748 8580] and at Parks of Drumquhassle [4840 8672 to 4848 8662]. The till, though usually brown in colour, at the latter site is reddish brown. Away from the influence of the reworked Linwood Formation clays, as at two localities in the Carnock Burn [4772 8313 and 4838 8370], the till becomes a more typical, variously coloured, diamicton composed of boulders, cobbles and pebbles in a sandy silty clay matrix. At the second site the red till incorporates blocks of tilted sand and gravel. The terminal moraine of the Loch Lomond Readvance Ice is largely composed of Gartocharn Till. South-west of Meikle Caldon [4878 8296] the moraine consists of soft red stony till with irregular gravelly patches, whereas to the north [4920 8337] it is described as being very sandy and bouldery. Sandstone erratics up to 7 m long have been seen in the moraine at Cameron Muir [4654 8286].

The Drumbeg Formation is named from the Drumbeg Sand and Gravel Pit near Drymen [484 882] (Sheet 38W), in which temporary sections have been visible for many years. The standard section is in the Gartness Borehole (Figure 27) below the topsoil (0.5–10.17 m). The two typical lithological associations here are fine- to medium-grained sand with silt layers, and silt with sand and clay layers. The sediments are reddish brown or orange in colour. Pebbly layers are present at the top as are rare beds of diamicton up to 4 cm thick. The sediments are mainly flat bedded, but locally are cross-bedded in the upper part. The presence of glacially induced faulting results in bedding dips well in excess of 30°. The deposit is in two upward-coarsening units. The base of the upper unit is at a depth of 6.4 m; the base of the lower unit is cut off by faulting. Where faulting is present, any adjacent beds of clay are usually sheared and contain striated polished surfaces.

Temporary sections in Drumbeg Quarry have revealed a wider range of lithological associations than in the Gartness Borehole [4973 8672]. This is to be expected as the quarry is located in a more proximal (upstream) position in relation to the source of the sediments. In the quarry, framework-supported gravel is present, in places crudely bedded but commonly trough and planar cross-bedded. However, sands are still the dominant lithology with trough cross-bedding more common than planar cross-bedding. On the eastern side of the quarry, and also in deeper excavations, horizontally laminated and ripple-laminated fine-grained sands (with silts) are much

Figure 27 Sections of the Quaternary deposits in the Mains of Kilmaronock, Gartness and Killearn boreholes.

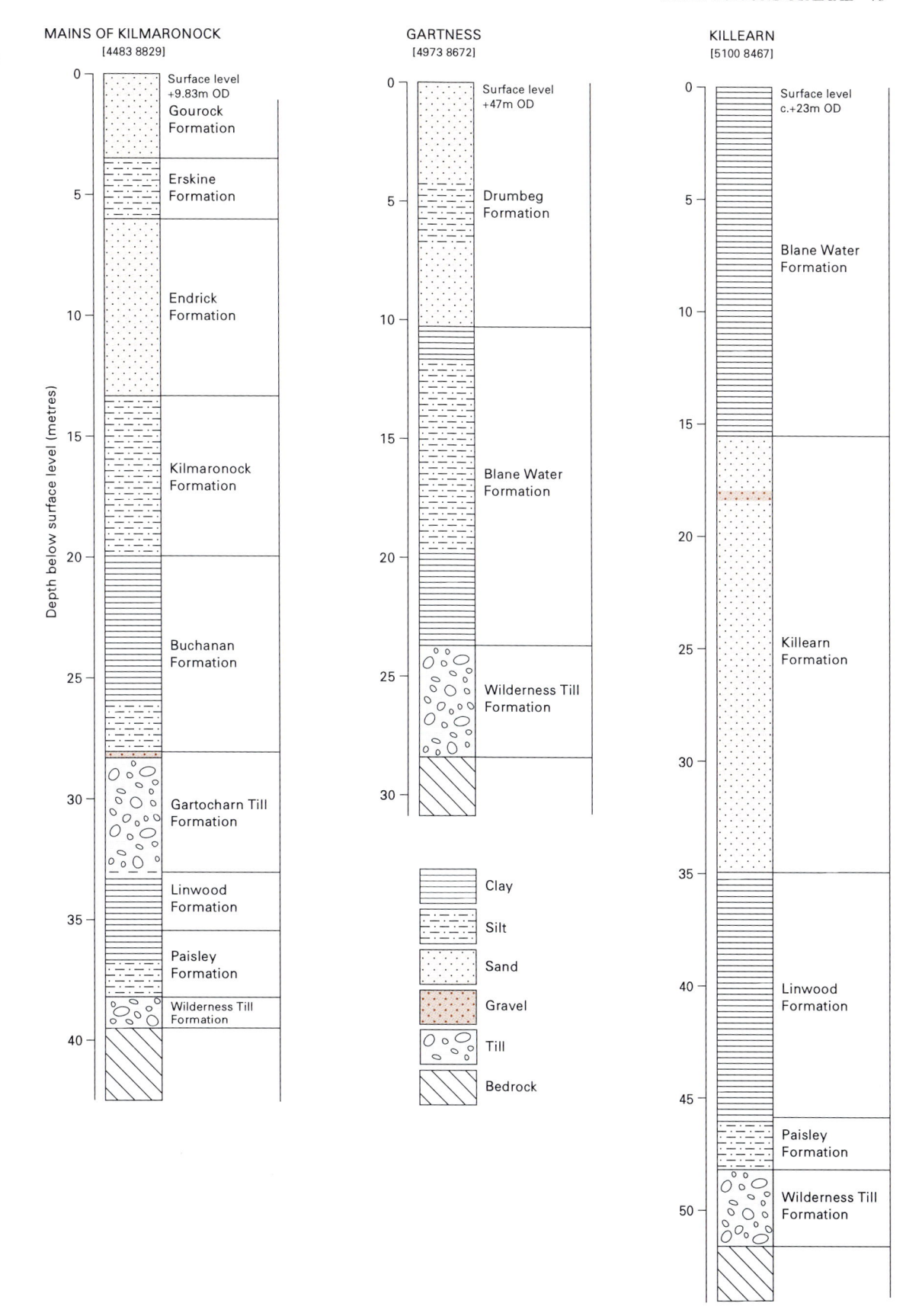

more common. Climbing ripples rise in an eastward direction, as is to be expected, consistent with transport directions indicated by the axes of trough cross-bedded units. Very little faulting of the sediments has been observed except in the most north-westerly exposures visible during 1992. This contrasts with the evidence of ice-pushing in the Gartness Borehole. Some of the gravel units are rich in marine shells and shell debris including the exclusively cold-water molluscs *Boreotrophon clathratus*, *Natica clausa* and *Chlamys islandica*. In the uppermost parts of the succession units of sandy diamicton are present; these may be solifluction deposits but might be better interpreted as debris flows.

Shells have also been reported from a 4 m-deep former gravel pit at Easter Catter [4743 8707] and in a railway cutting (Jamieson, 1865) at Upper Gartness [4940 8663]. In another section at Easter Catter [4783 8686] clay-bound gravel lenses and pods of diamicton are present in over 8 m of fine- to medium-grained sand. At Easter Drumquhassle [4886 8713] interbedded sand and silt with rare gravel lenses contains lobes of flow till. Slump folds and faults were also observed in this section which is located near the base of a steeply sloping north–south-aligned morainic ridge. Sections on another ridge farther east to the south-east of Gaidrew show red Gartocharn Till juxtaposed to sand, silt and gravel [4898 8626], contorted thickly bedded sand and silt [4904 8627] and shelly till [4906 8645]. At Upper Gartness, sand and pebbly sand are overlain by jumbled till and laminated silt and sand [4934 8666] whilst nearby contorted and faulted Blane Water Formation silt and red ?Wilderness Till are seen at Dalnair [4926 8662]. The structures and relationships seen in these sections clearly reflect the dynamics of an active glacier terminating in a lake. A small area of reddish brown pebbly sand in the Cameron Burn on Gallangad Muir [4587 8215] has also been assigned to the Drumbeg Formation. The bedding was described as of 'deltaic' dip. The deposit seems to have accumulated in a little lake dammed between the ice front and the head of the valley to the south.

The lithofacies associations of the Drumbeg Formation are typical of a deltaic environment of deposition in contact with active and decaying glacier ice. The presence of marine fossils in the Drumbeg area is misleading as they are believed to have been derived from older units such as the Linwood Formation. All the shells recovered are durable and show signs of abrasion as is to be expected from the milling processes taking place in the transport of the gravel. The Drumbeg Quarry deposits are lacustrine in origin and interdigitate with the Blane Water Formation clays and silts.

The Killearn Borehole contains the standard section (Figure 27) for the Blane Water Formation (0.5–15.61 m) (Plate 11). The typical lithology is of regularly interbedded clay and silt forming varves with dark coloured tops. The clays are silty, reddish brown and brownish grey, with grey silt, some sand layers and wisps, and thin beds of diamicton up to 7 cm thick with isolated stones up to 2 cm (interpreted as dropstones). Slumped bedding, flame structures and some microfaulting are present. A varve count yielded a total of 364 couplets in 11.45 m of preserved core (average couplet thickness 31 mm). If these figures are applied to the full 15.11 m of the formation at Killearn, the sequence represents some 480 years of deposition at one couplet per year.

A further section of the formation was encountered in the Gartness Borehole from 10.17 to 23.55 m. The clays and silts are much faulted (low- and high-angle reverse faults and antithetic normal faults), and have fine-scale sheared bedding with polished surfaces in places. Both features were produced by glacial disturbance, glacier ice having pushed into the deposits from the west. Rare rip-up clasts, sand beds, overfolding associated with thrusting, and primary current lineation were also noted. A varve count was not possible because of the faulting and because of the presence of multiple silt units.

The Blane Water Formation has also been seen at a number of field localities including the classic Croftamie site [4727 8608] described by MacFarlane (1858), Jack (1875) and Rose et al. (1988). Here the formation is sandwiched between the older Wilderness Till and the younger overriding Gartocharn Till. Felted organic detritus at the base of the Blane Water Formation has been radiometrically dated to 10 560 BP (Rose et al., 1988). Other localities demonstrate the complex relationships between formations which exist at and near ice fronts and terminal moraines. At Parks of Drumquhassle [4840 8672 to 4848 8662], shelly Gartocharn Till is overlain by Blane Water silt and clay which must have been deposited just as the Loch Lomond glacier had begun to retreat. However, in the banks of the Endrick at Park of Drumquhassle [4800 8709], typically the silt and clay rest on till (0.2 m) and also gravel (0.5 m+). Also by the Endrick, to the ESE of Dalnair [4992 8592], contorted silt with tilloid seams is seen to rest on Gartocharn Till beneath which is seen shelly clay (Linwood Formation).

The lithological associations developed in the formation are consistent with deposition of muds on a lake bottom in contact with, and near to, glacier ice. A cold climate at the time of deposition is indicated by the largely unfossiliferous nature of the sediments, the lack of organic content and the varved structure. The increase in the number of silt layers preserved in the succession at Gartness is probably due to the proximal position of the sediments in this borehole, in contrast to the coarse deposits of the laterally equivalent Drumbeg Formation just to the west.

The former extent of this glacial lake is not evident from the few shoreline features recognised. However, the distribution of the highest occurrences of the Blane Water clay and silt and related overlying Drumbeg sand and gravel appear to place limits to the lake's extent. For example, the clay reaches 55 m above OD at Station Wood, Killearn [5140 8576] and 60 m above OD in a section south-west of Carbeth House [5235 8719] where 0.9 m of clay rests on 1.0 m of sand and gravel and over 0.8 m of Wilderness Till. Farther south-east in Strath Blane, the clay has been found at over 45 m above OD at Levern Towers [5467 8035] and 47 m above OD at East Arlehaven [5429 8026]. Associated flat-topped deposits of sand and gravel are present hereabouts at between 50 m and 54 m above OD. However, around Killearn

Plate 11
Core from the Killearn Borehole.

a Seasonally laminated lacustrine sediments of the Blane Water Formation (12.92 to 14.87 m).

b Glaciomarine sediments of the Paisley Formation (46.31 to 47.75 m).

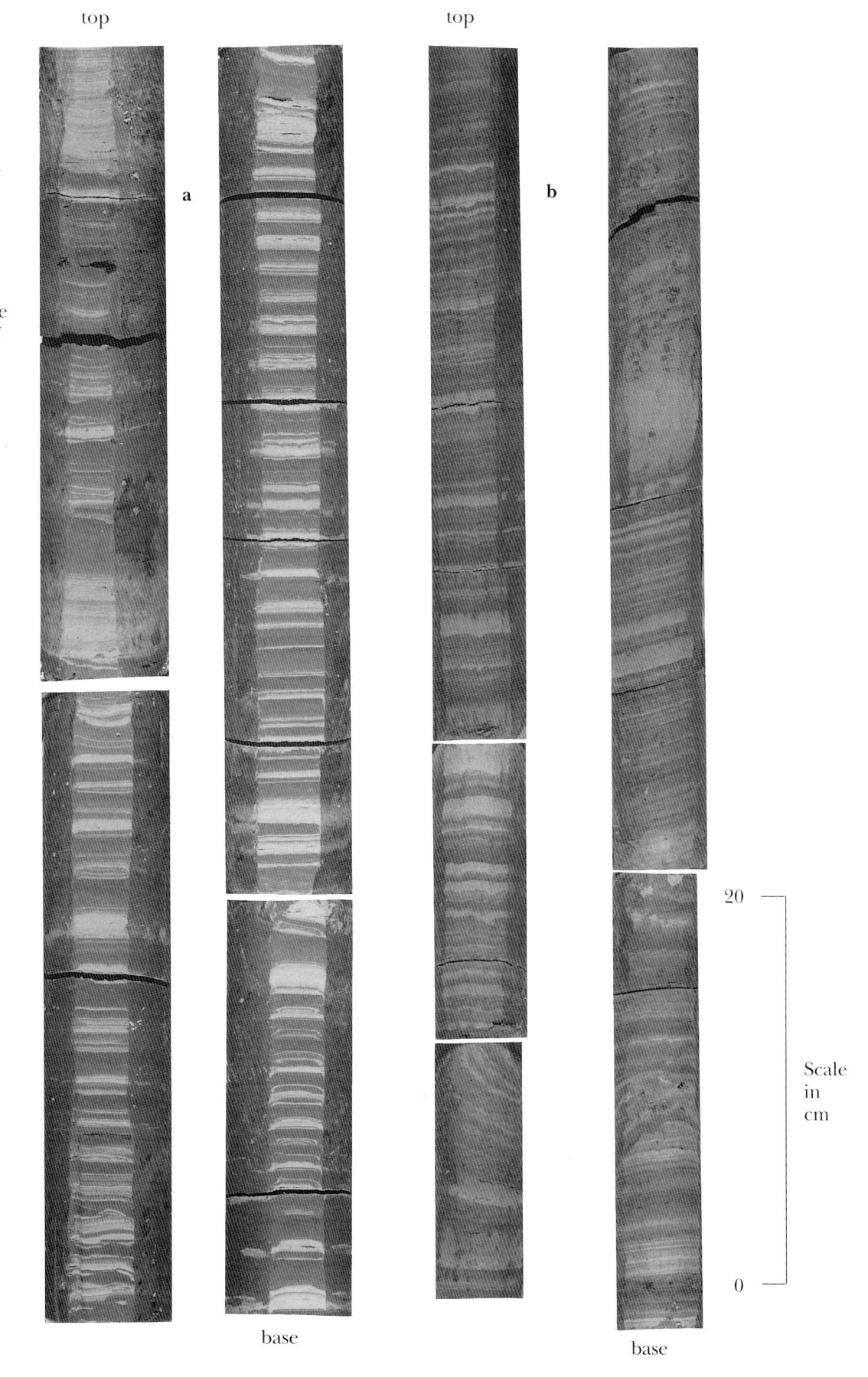

such deposits are known at higher levels between 65 m and 70 m above OD. From these or similar data and inferences it has long been recognised that the glacial lake in the Endrick valley discharged meltwaters to the Forth valley through the Ballat–Stoneyacre gap on Sheet 38E (Rose, 1981, fig. 6).

The Inverleven Formation is named from the Inverleven Borehole [3975 7509] at Dumbarton (Sheet 30W), which contains the standard section. The typical lithology is angular to subrounded boulders, cobbles, gravel and pebbles in a clayey sand matrix. The clasts appear to have had barnacles growing on them, although no attached specimens were retrieved. The deposit contains a marine fauna (Appendix 6). In the Clyde valley, the formation is represented in the Erskine Bridge Borehole by a shelly gravel at 7.5 to 10.02 m. The basal contact of this deposit probably represents a significant depositional break.

Several features of the Inverleven Formation led Browne and McMillan (1989a) to suggest that it formed as a lag gravel during a phase of tidally generated erosion, perhaps modified during a later phase of slow quiet deposition. Firstly, the matrix is very poorly sorted. Secondly, there is a marked spread of radiometric ages: paired valves of *Chlamys islandica* in the Inverleven Borehole were dated at 12 360 BP, whereas barnacle plates from here gave a radiometric age of 10 350 BP and from Balloch of 10 920 BP. The shells of *Chlamys* are well preserved, suggesting that transport during erosion was minimal.

The Inverleven Borehole contains the standard section for the Balloch Formation (8.5–54.0 m). There are three lithofacies essentially of clayey silt, silt, and clay. The deposit is colour banded in shades of mid to dark grey, brownish grey and sometimes reddish brown and contains rare marine shells (typically *Portlandia arctica*, and plant remains. Isolated clasts up to 2 cm (6 cm at Balloch), probably dropstones, are present as are normal and reverse-graded units usually less than 5 cm thick. Massive beds with degassing cavities occur in places as do traces of primary current lineation and slumped bedding. This formation possibly is present under the Clyde estuary in the district. The lithological and faunal characteristics of its clays and silts are consistent with deposition of muds on the sea bottom in front of and in association with advancing glacier ice.

At the time when Browne and Graham (1981) suggested that glaciation of this part of the district occurred late in the Loch Lomond Stadial it was contrary to accepted belief (Rose et al., 1988). However, Rose et al. re-examined the classic site at Croftamie [4727 8608] and have radiocarbon-dated an organic unit that lies below the Gartocharn Till and rests on Blane Water Formation muds. The age obtained, 10 560 BP, strongly supports the view that the local glaciation came late in the Loch Lomond Stadial.

Sea levels

Movements of relative sea level during the stadial are not well known. At or just before the time of the maximum extent of the ice, sea level may have been as high as 11 m OD, based upon the presence of former intertidal platforms and clifflines at about this height in the Firth of Clyde and around the shores of Loch Lomond (Rose *in* Jardine, 1980 pp.29–31). The view that such features were formed by severe shore erosion in the periglacial environment of the Loch Lomond Stadial (references in Jardine, 1986) must be questioned in the light of the discovery of till on the platform at Cardross in the Greenock district. The evidence presented by Browne and McMillan (1984) implies reoccupation of a pre-existing marine feature here, and by implication elsewhere, during the Loch Lomond Stadial. In Danes Drive, Scotstoun [532 679] and at Shiels, near Renfrew [523 665] drumlins, planated by the sea in Loch Lomond Stadial times, now form parts of the former intertidal platform as developed in the Clyde valley. Former excavations in the Glasgow Road, Clydebank [4990 6987 to 5055 6925] showed potholes up to 7 m deep eroded into the Wilderness Till which here forms the platform. Carruthers' notes describe the infill as mud and water. The presence of potholes might not be expected if the platform was entirely cut in periglacial conditions.

On the east coast of Scotland sea level may have fallen below that of the present at some period during the Loch Lomond Stadial (Browne, 1985 in discussion of Sutherland, 1984; Armstrong et al., 1985, pp.87–89). The implications are that on the west coast sea level was also lower than OD. This is a possibility in the Clyde as implied by McGown and Miller (1984, fig. 7) for central Glasgow, where they recognise a channel incised to over 40 m below OD, aligned close to the course of the River Clyde and backfilled with river sands and gravels. However, dating of the supposed incision is open to doubt because of the problem of correlating largely unfossiliferous superimposed successions of sands and gravels which could include the late-Devensian Broomhouse, Killearn and Bridgeton formations (Figure 20) overlain by similar Flandrian formations. The incision may therefore, as previously suggested, be related to much earlier erosional events.

Landslip and solifluicted deposits

Landslips are common on the upper slopes of the Campsie and Kilpatrick hills (Plate 12). They were probably initiated during the Loch Lomond Stadial. They consist principally of rotated blocks of basaltic lavas and tuffs (Plate 13). Adjacent areas of thin gelifluicted drift (head) are also present. The largest landslips which cover an area of up to 1 km^2 occur beneath basalt lava scarps near Lang Craigs [430 765], Tomibeg [510 773] and at the corries of Balglass [580 855 to 590 855]. Other extensive areas are found north of Blanefield [55 85] and on the Kilpatrick Braes [474 790].

In Glasgow unusual subsoil profiles observed on the flanks of drumlins may also be the product of large-scale solifluction (Dickson et al., 1976). Both at Robroyston [633 676] in the Airdrie district and at Springburn [609 678] (Forsyth et al., 1996, fig. 24) a grey unweathered till overlies peat, which is radiometrically dated at between 11 650 and 11 100 BP. The peat contains a flora indicative

Plate 12 Landslip and scree on the flanks of Dumgoyne, a volcanic vent of Lower Carboniferous age, near Strathblane (C 2108).

Plate 13 The Whangie, landslipped mass of basalt from Lower Carboniferous intrusion, showing deep, open back fissure, Auchineden Hill (D 122).

of tundra conditions. At Springburn the till is overlain by a younger peat dated at 5994 BP. Because these sites lie well beyond the limit of the Loch Lomond Stadial ice, the tills seem most likely to be the products of the Dimlington Stadial glaciation. The available evidence therefore suggests that till has moved over the organic materials probably by solifluction processes which were a major feature of periglacial activity in west-central Scotland during the Loch Lomond Stadial. At least two units of soliflucted sediments were recorded in the section described by Mitchell (1952) at Garscadden Mains [5341 7124]; the lower was of Loch Lomond Stadial age.

Soliflucted deposits are probably present above the marine succession in the Inchinnan area. A section seen in a sewer trench [4764 6935] showed 2.7 m of thinly bedded, brownish grey clayey silt, with dark grey laminae and sand layers. Layers, 2 to 4 cm-thick with plant debris, were present at the base and 14 cm above. This silt rested on a brownish grey, clayey in part, gravelly, sandy, diamicton 0.9 m thick. The gravel portion was subangular to angular in the lower part, and subrounded to rounded in the upper part, the clast size averaging 2 to 4 cm. The abundant matrix was massive and contained marine shells. Underlying the diamicton is shelly, brownish grey silty clay that is also gravelly with angular to subrounded clasts up to 5 cm. This deposit is part of the Linwood Formation. As the locality is well outside the limits of the Loch Lomond Readvance Ice and shells from underlying sediments nearby are under 13 000 years old, the diamicton in this section is unlikely to be glacial in origin. The most likely explanation for its origin is that it is a marginal marine gelifluction deposit of Loch Lomond Stadial age. The overlying silt is interpreted as an alluvial sediment interlayered with solifluicted material partly infilling the floor of the minor valley where the section is located.

At Portnauld Farm railway cutting [4915 6855 to 4940 6857] a possibly similar gelifluction deposit was seen by Jack. Described in his notebook of 1870 as 'unmistakenly boulder clay', the diamicton had the appearance of a coarse gravel especially near the top. The stones were enclosed in a stiff unstratified paste, were lying at all angles, and many were distinctly striated. The bed was 0 to 0.9 m thick. Beneath, was a stony, sometimes clayey sand. Shells occurred in groups, with balanids adhering to the stones and on some shells. Paired valves were common. Some of the stones were large and considered to be washed out of the underlying till. The bed had an undulating base and was up to 0.3 m thick.

Further occurences of solifluicted diamictons may also explain the presence of apparently typical Linwood Formation shelly deposits below 'boulder clay'. Unlike Inchinnan and Portnauld which are both lower than 12 m above OD, these localities (seen by Jack) are higher, approaching the late-Devensian marine limit as recognised in the Linwood–Paisley embayment. In the first section [4112 6482] at Tweeniehills railway cutting (SL c.36–37 m above OD), the surface deposit is a diamicton composed mostly of shale fragments. Some of the small stones are distinctly striated but there is no sign of stratification. This deposit was said to be similar to that at the top of a section at Windyhill. It may be 3.4 m thick, resting on a dark blue, finely laminated clay that is rather sandy in places. Shells were noted only in the sandy part of the succession. The clay is 1.4 m thick but dies out eastwards within 6 m. Underneath is another diamicton at least 1.7 m thick and believed to be a glacial till. In the eastern section [4118 6478] in the cutting (SL c.34 m above OD), a blue clay up to 0.6 m thick with a thin peat layer is underlain by a very stiff, laminated, blue clay with very fragmentary mussel and other shells in a layer 1.2 m from the surface. This clay was about 1.2 m thick and rested on a diamicton 1.4 m thick. No trace of the solifluicted diamicton was recorded.

In the Windyhill railway cutting [4220 6455] (SL c.27 m above OD), an unstratified diamicton 1.2 m thick, composed of debris of shale, contains a few fragments large enough to show that they were glacially striated. Underneath is a stratified blue clay, 0.45 m thick, with occasional layers almost wholly of Carboniferous shale fragments. This deposit rested on another blue clay (1.2 m thick) that included a shell bed (0.30 m thick). At the base of the section was another glacial diamicton 1.1 m thick.

A diamicton at surface at Drumcross Farm [4493 7120] may also be a solifluction deposit. The locality is at about 27 m above OD. The diamicton (described as boulder clay) had stones of all sizes up to 36 cm lying in all directions. The stones were not rounded or striated. The bed lacked stratification and was up to 1.22 m thick. Beneath was a stiff clay with shells, much 'interrupted' by stones of the same shape as in the diamicton above. There was a very large stone at the top with balanids adhering above and below. The clay had sandy pockets and became very sandy and even gravelly towards the base. Specimens of *Mya truncata* were seen by Bennie and Jack. This clay bed (up to 36 cm thick) is believed to be lenticular and very localised as the adjacent section, reportedly recorded by Fraser, was of a 15 cm-thick, lenticular, shelly gravel.

One further locality at which solifluicted diamicton may be present is on the Bargaran and Shilton March Burn [4578 7089]. The surface level is around 26 m above OD. Here a diamicton (described as boulder clay), of unknown thickness, rests on a strong, dark blue clay containing driftwood (Scots pine), fragmentary decayed shells and a beetle.

FLANDRIAN

The Loch Lomond Stadial ended about 10 000 years ago and the glacier ice in the district and around Loch Lomond seems to have disappeared quickly. The succeeding warm phase, the Flandrian Stage, is marked by marine and nonmarine deposits and landforms in the Clyde area. An area of controversy still unresolved is the relationship between the raised peats of the Paisley–Linwood area and the extent of marine transgressions across and around them in Flandrian times. From offshore cores in Loch Lomond (Dickson et al., 1978) and the data from the Mains of Kilmaronock and Linwood boreholes (Boyd, 1982; Elliott, 1984) the vegetational history of the Flandrian is reasonably well known.

Open heathland that characterised the late-Devensian was re-established after the Loch Lomond Stade glaciation. This was replaced by landscapes dominated by woodlands of hazel (*Corylus*) from about 9000 years ago, by a major increase in alder (*Alnus*) around 7000 years ago and a major decrease in elm (*Ulmus*) about 5000 years ago. However, the correlation of these major vegetational changes with radiocarbon-dated successions at Linwood and in Loch Lomond allows an appraisal to be made of the available evidence relating to Flandrian sea-level change in the district and marine transgressions into the Loch Lomond and Strath Blane basins.

Strath Blane basin

The Mains of Kilmaronock Borehole [4483 8829] provides evidence for three possible marine incursions into the Strath Blane basin from Loch Lomond during the early to middle Flandrian. The earliest incursion is represented by the seasonally banded muds of the Buchanan Formation (Browne and McMillan, 1989a) and is recognised on the basis of the occurrence of dinoflagellate cysts and foraminifera. This formation rests almost directly on the Gartocharn Till Formation, here consisting of glacially reworked marine muds from the Linwood

Formation. As a result, there is a possibility of faunal reworking from the older formation. The Buchanan Formation is over 9000 years old, as the pollen analysis indicates the presence of pioneer heath vegetation. Sea water could have entered Loch Lomond from the Vale of Leven at times of high sea level, as was the case at that time in the Forth (Sissons, 1976).

The typical lithology of the Buchanan Formation is a thinly bedded silty clay with many laminae and thin bands of silt and sometimes of sand. The deposits are brown, brownish grey and reddish brown, and are conspicuously colour banded, the banding being emphasised by the grey coloured units of clay. Isolated clasts up to 4 cm are present (possibly dropstones). Overall the dark and light colour banding of the deposit suggests seasonally controlled deposition (varves); the thickness of the couplets decreases upwards from as much as 15 cm to less than 1 cm.

The Buchanan Formation is overlain by the lacustrine muds of the Kilmaronock Formation. The Mains of Kilmaronock Borehole contains the standard section (13.60–19.95 m). The typical lithology is a very thinly bedded silt with many clayey silt and silty clay layers. Sand layers are also common. The deposits are brown, dark brown, brownish grey and grey in colour, with plant remains and locally with dark organic-rich bands. Grains and flecks of vivianite are present throughout. The lithology of the Kilmaronock Formation is consistent with deposition of lake bottom muds in the Loch Lomond basin and perhaps the Strath Blane basin. The alternation of organic-rich and organic-poor horizons suggests seasonal controls on deposition. The increasing abundance of *Corylus* pollen in the upper half of the formation would suggest an age of not more than 9000 years (Price, 1983).

The Kilmaronock Formation is overlain by the predominantly upward-coarsening sandy Endrick Formation of deltaic and fluvial origin (5.99–13.60 m in the Mains of Kilmaronock Borehole). The typical lithology is loose, fine- to medium-grained sand with some silt layers. The deposit is reddish brown and contains abundant plant remains and some dark organic clay bands.

A second marine incursion possibly took place over 7000 years ago (pre-*Alnus* rise). It is represented by a bed of grey silt with dinoflagellate cysts in the Endrick Formation at a depth of 9 m in the Mains of Kilmaronock Borehole. At the top of the formation a mixed forest flora is indicated by the pollen and, following Price (1983), this would mean that the strata at about 6.6 m depth in the borehole are at least 6500 years old.

The stratigraphical record of the Mains of Kilmaronock Borehole is completed by a third marine incursion represented by the Erskine and Gourock formations. Not only are dinoflagellate cysts present in the Erskine Formation but there are at least two layers composed of partly rotted shells of the bivalve *Mytilus edulis* (at depths of 5.15 m and 5.46 m respectively). As there is no indication of the *Ulmus* decline in the pollen record this unit is at least 5000 years old. Based upon the report of Dickson et al. (1978) on almost certainly the same marine incursion found in offshore cores, the transgression lasted from around 6900 to 5400 years ago.

Lower Clyde valley

In the Linwood and Paisley embayment of the lower Clyde valley, two views of the extent of the Flandrian marine incursions have been proposed. Browne (*in* Jardine, 1980) suggested that the large peat mosses may have excluded the sea from most of this area and that extensive flat tracts at about 6 to 9 m above OD are, for the most part, remnants of a late-Devensian surface produced by erosion of slightly older late-Devensian estuarine sediments. The surface concerned may have been produced at the time of the 'Main Late-glacial Shoreline'. Boyd (1982, fig. 8:4) suggested a more extensive area of Flandrian submergence, on the basis of a detailed study of the Linwood Moss area. He identified a mid-Flandrian shoreline position, with associated deposits of silt up to 6 m thick and extending up to 8 m above OD. Boyd also identified an 'old' and a 'young' peat. The former is present in his Linwood Moss Wood Borehole [4395 6605] and in the BGS Linwood Borehole, the latter in his Moss Cottage Borehole [4440 6610]. There is doubt about the apparently young age of 9290 BP (Boyd, 1986) for the base of the 'old' peat obtained in the Linwood Moss Wood Borehole. In the Linwood Borehole the base was dated at 9540 BP. The base of the 'young' peat was dated at 3650 BP in the Moss Cottage Borehole. These dates are consistent with previous ones reported from this area (Boyd, 1982; 1986). From the radiocarbon dates alone, it is obvious that the history of Linwood Moss is complicated. No research has been carried out on the histories of the nearby Barochan, Fulwood and Paisley mosses or on remnant patches of peat in the lower Clyde valley east of Renfrew.

The only stratigraphical evidence for Flandrian marine transgression in the Linwood–Paisley area is a 5 cm thick band of grey mud within the Clippens Peat Formation in the Linwood Borehole. This is unfossiliferous apart from a specimen of *Onoba semicostata*, a marine gastropod which may have been derived from late-Devensian sediments. However, taken with pollen evidence of a hydroseral influence in the peat below and above the mud, an estuarine or near-estuarine origin for the deposit is acceptable. The mud unit is probably younger than the 8000 BP that Boyd (1982) envisaged, for a radiocarbon assay of the base of the overlying peat gave an age of 7110 BP. However, this marine event, which ended before the *Alnus* rise, is older than the middle Flandrian marine transgression (6900–5400 BP) that is represented by the Gourock and Erskine formations in the Loch Lomond basin. However, the 7110-year date and *Corylus*-dominated pollen record do suggest a correlation of the *Onoba*-bearing mud with the dinoflagellate cyst-bearing grey silt at 9.4 m in the Mains of Kilmaronock Borehole. Although there are reservations about identifying these two deposits as marine, on the basis of a single shell and a few dinoflagellate cysts, it is considered that both relate to the same rise of relative sea level, on the evidence of their similar ages.

Evidence for changes in Flandrian sea level is seen in the presence of three series of Flandrian to Recent levels in the lower Clyde valley (Figure 28). The upper limits of

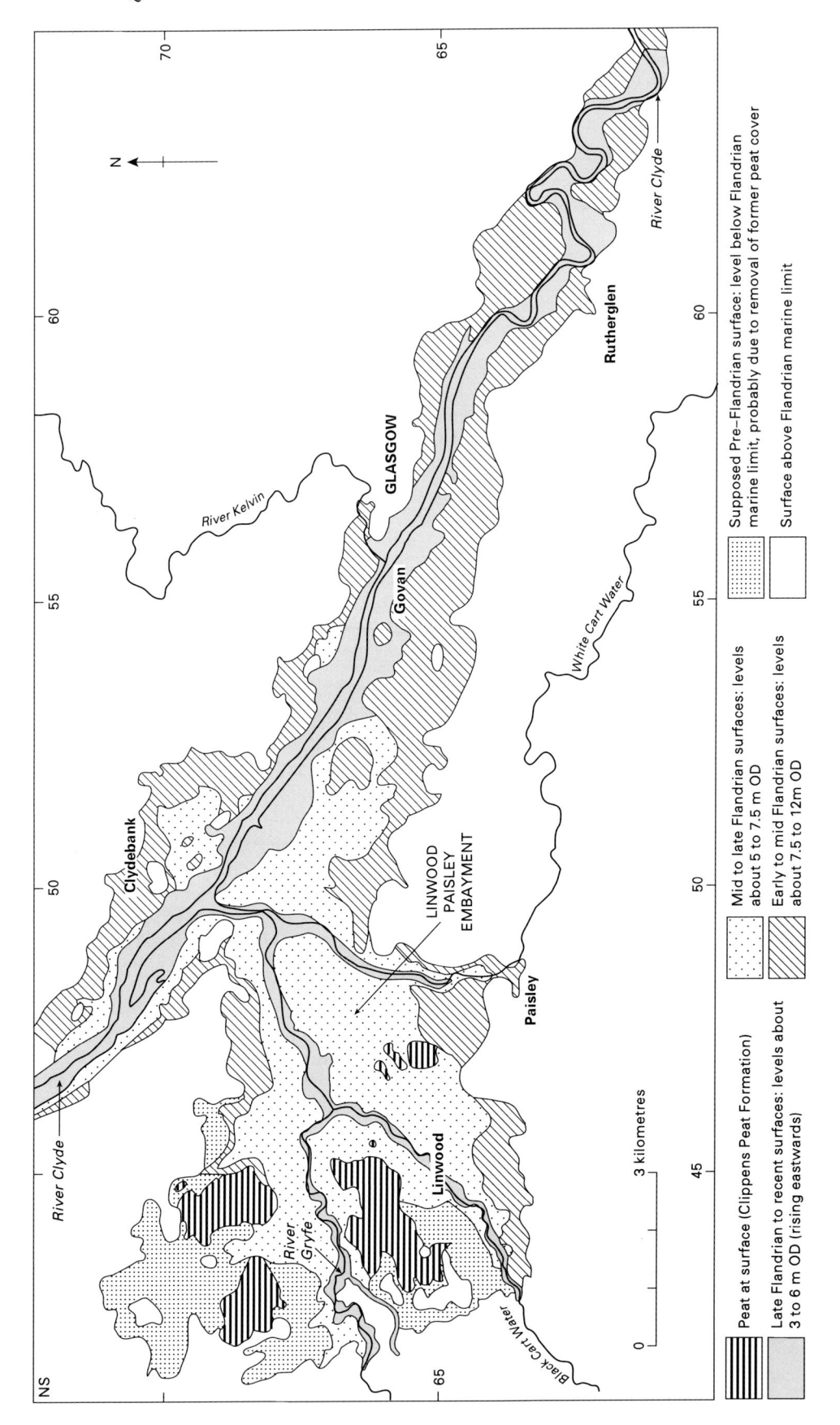

Figure 28 Generalised distribution of Flandrian estuarine flats in the lower Clyde estuary.

these surfaces are at about 12 m above OD, less than 7.5 m above OD and less than 5 m above OD (based on a review of Ordnance Survey spot heights). However, in the Linwood–Paisley embayment only the lowest of these levels can be recognised. Between 5 and 12 m above OD there are generally gently sloping surfaces not divisible into discrete terraces, and these are overlain by remnants of previously more extensive peat mosses. In both areas there is generally a marked, low cliffline at about 12 m above OD. This feature is partly concealed by peat to the west of Barochan Moss, north of Fulwood Moss and south-west of Linwood Moss. This fact suggests that the cliffline could be an inherited feature of late-Devensian age, certainly in those areas where the mosses are likely to have accumulated continuously from the beginning of the Flandrian. From Boyd's work (1982) it is also likely that, wherever the surface level of the moss approaches 15 m above OD, peat accumulation has been continuous from the beginning of the Flandrian, and the surface and sediments below are of late-Devensian age. Indeed, the late-Devensian Linwood Formation is at surface over extensive parts of the flat ground between 8 and 12 m above OD in the Linwood–Paisley area.

However, it appears that the Flandrian marine incursion flooded all that part of the Linwood–Paisley area below 7.5 m above OD, the general upper limit of the middle series of surfaces in the Glasgow district. Above this limit, the only evidence for Flandrian transgression is the mud unit with the *Onoba* shell in the Clippens Peat Formation in the Linwood Borehole at about 8.4 m above OD. Boyd (1982, fig. 8.4) constructed his conjectural configuration based on an altitudinal limit of 8.7 m above OD outside the mosses and a surface level of 11 m above OD inside. The reconstruction shown in Figure 28 is based on similar premises.

The deposition of the 'young' peat marks the final regression of the Flandrian sea in the district. The Moss Cottage radiocarbon date of 3650 BP for the base of the peat places sea level well below 7.8 m above OD by this time. In fact, it is likely that sea level was below 5 m above OD, this level being the lowest recorded on the outer edge of the middle series of estuarine flats in the valley.

Estuarine deposits

The oldest Flandrian estuarine deposit is the Longhaugh Formation. It consists mainly of grey, loose, fine- to medium-grained, sometimes silty sand containing some comminuted shell debris. It forms the infill of an erosive slot cut into older estuarine and glacial sediments in the lower Clyde valley. The floor of this channel appears to be 20 m or more below OD in central Glasgow (McGown and Miller, 1984, fig. 7). Gravel with shells was recorded in the Longhaugh No. 20 Borehole [4291 7319], below sand, between levels of 19.90 and 21.26 m below OD. The Longhaugh Formation was probably deposited in estuarine channel environments.

The Erskine Formation (Figure 27) comprises subtidal to intertidal silty clays with sand and silt laminae and beds. Organic debris and isolated pebbles are present. The deposit is generally brownish grey or greyish brown but black where it contains much iron sulphide. The sand and silt is ripple laminated in places.

The Gourock Formation consists predominantly of sand, commonly with gravel. It was proved in the top part of the Bridgeton Borehole where the uppermost metre consisted of brown silt and clay with thin beds and wisps of peat and sand. The Bridgeton sequence is typical of this part of Glasgow as the fieldnotes of Jack and Bennie describe several sewer trenches displaying similar sediments [6080 6375, 6078 6411, 6083 6357]. Elsewhere in the district, sections indicate that the direction of transport is consistently towards the west. Dark grey beds rich in organic detritus may be present as in the Shieldhall sections [535 663]. They are also present in the Dalmarnock area as in Dale Street [6083 6357] where, under 1 m of clay, Jack and Bennie recorded 2.1 m of dark river silt with oak driftwood, leaf beds and vegetable debris, these sediments resting on at least 1.3 m of sand and gravel. In John Street [6048 6376] a similar thickness of silt was recorded as containing grasses, sedges and acorns; the silt also had diatoms in it at Muslim Street [6093 6373]. In Hutchesontown [5987 6290] they also noted the presence of an angular block (1.3 m) of laminated clay (i.e. the Paisley Formation) largely encased in sand and gravel but covered by yellow loam (i.e. the clay of the above Dalmarnock localities).

At Windmillcroft Dock [583 647] sandy gravel contains boulders of glacial till, pebbles of brickclay and peat, artefacts and periostraca of the freshwater mussel *Unio margaritifera*. Plant remains are common including a large trunk of an oak but no beech. The formation has a markedly erosive base, and locally is up to 5.2 m thick where it infilled a channel cut into the underlying units.

The lithological association of the Gourock Formation is compatible with an estuarine origin for the sediments. However, the estuarine environments represented range from one dominated by fluvial processes in the east, in Glasgow, to an almost fully marine environment in the west, at Gourock. The fluvially dominated sediments were deposited in very shallow channels probably linked in a branching anastomosing pattern, isolating low-lying islands in the estuary. In the marine-dominated part of the estuary a pattern of shallow channels forming a complex with extensive intertidal flats is likely.

Peat deposits

Peat, radiocarbon-dated as pre-Flandrian in age, has been found in a few places in the Glasgow district. As mentioned above (Landslips and solifluted deposits), examples are known from Springburn and from Robroyston in the Airdrie district (Dickson et al., 1976). At Springburn [609 678] geliflucted till overlies peat dated at 11 140 BP and is overlain by a thin Flandrian peat dated at 5994 BP.

The principal deposits of peat found in the district are of Flandrian age. During the early Flandrian, as the Scottish climate became warmer and wetter, *Sphagnum* peat developed in hollows on former lake floors and on poorly drained flat areas. Blanket bogs also formed on sloping ground in the Campsie Fells and Kilpatrick Hills in response to exceptionally high rainfall. Peat, which in the Glasgow area has been assigned to the Clippens Peat Formation (Browne and McMillan, 1989a, pp.35–36), formed in low lying ground particularly to the north of Linwood. Here extensive deposits have accumulated as raised mosses.

The Linwood Borehole contains the type section of the Clippens Peat Formation. Below 45 cm of very peaty stony soil, 1.3 m of peat rests on 5 cm of grey rooty clay, followed by a further 84 cm of peat to the base of the formation at 2.64 m depth. The surface level of the borehole was 10.18 m above OD; thus the base of the upper bed of peat was at 8.43 m above OD and the base of the lower at 7.54 m above OD. Samples representing the basal 3 cm of each peat bed have been radiocarbon dated. The ages are 7110 BP for the base of the upper peat and 9540 BP for the base of the lower bed. Boyd (1982; 1986) recognised that *Calluna* and *Ericales* have peaks in their pollen curves (16 per cent and 18 per cent respectively) on either side of the 5 cm clay band at 1.75 to 1.80 m. These peaks may reflect more open conditions just before and after the clay was deposited. As noted above, a single specimen of the marine gastropod *Onoba semicostata* was found in the clay, tentatively indicating a marine origin for this bed, the transgression ending earlier than 7100 years ago. Reliable indicators of saltmarsh vegetation are not found in the peat record. However, the ericaceous pollen taxa may indicate the development of a coastal heath. The pollen spectrum, as a whole, reflects an open environment which becomes progressively wetter and with an increasing number of trees at higher levels.

Fluvial and lacustrine sand and gravel

Coarse-grained Flandrian sediments deposited in fluvial environments including present-day river courses were assigned by Browne and McMillan (1989a, pp.39–40) to the Law Formation (BGS, 1994, sections C and D). Older Flandrian sand and gravel which may have been partly associated with deltaic and lacustrine environments were referred to the Endrick Formation (Browne andMcMillan, 1989a, pp.36–37).

The Law Formation includes much of the Recent sand and gravel alluvium of the River Clyde together with the Endrick Water, the Blane Water, the Black and White Cart waters, the River Gryfe, the River Kelvin and the Glazert Water. The lithology is of loose, grey, fine- to coarse-grained sand with some silt and beds of fine to coarse gravel. Plant remains are present and dark sulphide patches occur. The deposit may be partly massive, sometimes flat-laminated but probably mainly cross-bedded.

In the valley of the Glazert Water up to 10 m of coarse gravel, boulders and sand are present. The deposits are likely to have been derived from Strath Blane to the north-west via the wide 'dry' valley of the Pow Burn. The whole sequence has been assigned to the Law Formation but the basal part could have been deposited by glacial meltwaters during either of the late-Devensian glaciations. Deposits of the Endrick Formation may be present under 'alluvial' terraces of the River Kelvin lying above the general level of the floodplain at elevations of 35 to 50 m above OD.

Fluvial and lacustrine clay and silt

The Kelvin Formation includes the Recent fine-grained sediments of the valley floor of the River Kelvin and other rivers and streams. Lake alluvium of Flandrian age, present north of Glasgow and west of Bishopbriggs at elevations of about 40 m above OD, is also referred to this formation. The sediments are silty clays and silts, often with organic remains, interbedded with peat, generally formed either on the floodplain or in lacustrine environments such as inter-drumlin hollows. Clay and silt of the Kelvin Formation forms much of the valley floor of the River Kelvin (BGS, 1994, section C). Borehole records show the thickness of these sediments to range from about 2 to 9 m.

In the Kelvin valley lacustrine clay and silt deposits concealed by younger Flandrian sediments are locally present. These deposits probably belong to the Kilmaronock Formation (Browne and McMillan, 1989a, pp.35–36). Typically, the sediments comprise thinly bedded silt with clay laminae and some sand layers. Plant remains are common.

Man-made deposits

Extensive areas of man-made deposits are present in and around the urban areas of the Glasgow district. Wherever construction for houses, factories and roads has taken place there are likely to be areas where the natural ground surface has been covered by redistributed, geotechnically variable, natural and man-made materials. Typically made ground of this nature is only a few metres thick but in areas close to former excavations thicker deposits may be expected. Examples of major ground disturbance include the M8 motorway and interchanges such as that at Paisley [465 655]. Former pit bings and slag tips also constitute made ground as does the spoil from the dredging of the bed of the Clyde and the excavation of the Clyde Road Tunnel (Nicoll, 1990).

The district has been extensively quarried for minerals including sand and gravel, brick clay, hardrock aggregates, fireclay, common shale, quartz-conglomerate, sandstone and coal. Many former excavations are infilled or partially infilled with variably compacted materials and these deposits, classified as Fill on the Glasgow Drift Sheet, are generally thicker than other made ground and may be in the order of tens of metres.

TEN
Economic geology

In the past, the rocks of the Glasgow district have been extensively exploited for several minerals. At the present time (1994), however, only sources of aggregate and roadstone are economically viable and being exploited to any extent.

RESOURCES — SOLID

Coal

The Limestone Coal Formation is the principal source of coal in the district. At least 13 seams have been mined (Figure 11), some of them extensively, such as the Knightswood Gas and Possil Main coals. All but the Kilsyth Coking Coal are in the upper part of the formation. The Middle Coal Measures are rich in thick coal seams, six of which have been mined (Figure 14), but they only occur in a limited area in this district. The Lawmuir Formation has also produced a great deal of coal, partly from the Quarrelton Thick Coal at Johnstone and its component seams in and around Paisley. The Hurlet Coal has also been worked extensively both south and north of the River Clyde. The Lillie's Shale Coal is the only seam which has been exploited in the Lower Limestone Formation, mainly in the Linwood–Johnstone area. In the Darnley area and in north-west Glasgow, several coals in the Upper Limestone Formation have been mined but to limited extents. The Kiltongue and Airdrie Virtuewell coals in the Lower Coal Measures have been extracted in southern Glasgow to quite appreciable extents. Coal mining began over 300 years ago in Paisley and the Govanhill area of Glasgow. It ceased with the closure of Garscube Colliery [573 698] which finally became uneconomic in 1966.

Ironstone

Both blackband and clayband ironstones have been much exploited in the Glasgow district. The Limestone Coal Formation is the main repository for both types with at least six of the former and three seams of the latter known to have been worked. The Johnstone and Garibaldi clayband ironstones and the Lower and Upper Garscadden, Jordanhill and Lower and Upper Possil blackband ironstones were the most extensively extracted. The Johnstone Clayband Ironstone was mined almost entirely south of the Clyde, the Lower Garscadden (or Linwood) Ironstone extensively both south and north of the river and the others mainly north of it. Mining of the blackband ironstones began about 1830 and ended at Cadder No. 17 Pit [5949 7165] in 1923. Clayband ironstone working may have begun even earlier and ended at Victoria Pit [521 506] in 1921. The only other group to contain workable ironstones is the Lower Limestone Formation, which has the Lower and Upper Househill clayband ironstones in south-west Glasgow and the latter's equivalents, the Campsie Clayband Ironstones, north of the city. At least one attempt was made to work an ironstone of unknown type, possibly above the Castlehead Lower Coal, in the Lawmuir Formation south of Paisley but it was quickly abandoned over 100 years ago. Mining essentially became uneconomic when thick, low-grade deposits, such as the Mesozoic ironstones of England, became available.

Limestone

Several limestones in the Lawmuir Formation and in the Lower and Upper Limestone formations have been quarried and some have been mined as well. Most of the workings have been abandoned for over 100 years. The stratigraphically lowest is the Hollybush Limestone, which was quarried at Hollybush [495 612], Brownside [489 606], Brediland [461 629], Elderslie [447 631] and Garnieland [481 696], where it also appears to have been mined. The Blackbyre Limestone was quarried at Nethercraigs [467 610]. The Baldernock Limestone was quarried at Blackhall [497 628] and mined at the Linn of Baldernock [591 757], where the stoop-and-room workings are still visible. The Hurlet Limestone was quarried and mined in several places, for instance at Hurlet [NS 513 611], Nethercraigs, Duntocher [422 728], Windyhill [523 766] and on the South Brae of Campsie around Newlands [608 766], mostly as a combined seam with the Hurlet Coal and the Alum Shale, though not as extensively as the coal alone. In the Upper Limestone Formation, the Lyoncross Limestone was mined to a small extent at Waulkmill Glen [521 581]. The Orchard Limestone was quarried and to a small extent mined as a cementstone at its type locality [562 590]. The Calmy (or Arden) Limestone was extensively quarried and latterly mined at Darnley [525 589], where it is unusually thick (3–4 m). Operations did not cease until 1960 when the more accessible resources were worked out, making this much the most recent limestone working in the district.

Several beds of cornstone in the Kinnesswood Formation have been quarried on a small scale and burnt locally, the most extensive workings being on Dumbarton Muir [438 807].

Sandstone

Sandstones of Lower and Upper Devonian age have been quarried for building stone in the northern part of the district. The former was worked in several quarries near Gartocharn [43 86] where purple-red sandstone was extracted. The latter was worked at Croftamie [478 857] and from several quarries on Blairquhomrie Muir [430 825] exploiting bright red, easily dressed sandstone. Auchincarroch Quarry [423 819], just to the west of the district, had a working face about 50 m high.

Sandstones from the Lawmuir, Limestone Coal and Upper Limestone formations and the Upper Coal Measures [606 612] have also been quarried, in many cases for local use only. The Nitshill quarries, in two sandstones in the Limestone Coal Formation (Nitshill and a lower one) were quite extensive. However, the two most important sandstones in the district, the Bishopbriggs and the Giffnock, are both in the Upper Limestone Formation. Both were very extensively quarried and subsequently mined for building stone at their type localities, Bishopbriggs [609 695] and Giffnock [569 592]. They were abandoned as uneconomic early in the 20th - century, leaving galleries 15 m high.

Conglomerate

The Douglas Muir Quartz-Conglomerate Member, which occurs in the lower part of the Lawmuir Formation, crops out north and north-west of Glasgow. Currently, the deposit is being worked at Douglas Muir [526 748] and Strathblane [574 748]. The pebble content of the conglomerate is almost entirely vein quartz and, when disaggregated, produces a very low shrinkage aggregate and good clean sand suitable for glass and foundry work (Cameron et al., 1977).

Mudstone, oil shale and fireclay

Mudstones in the Lower Limestone Formation have been extensively quarried for brick clay around Blairskaith [595 755] and at Fluchter [588 747]. Several seams of fireclay in the Lawmuir Formation were mined at Paisley from the Ferguslie [4606 6346] and Caledonia [4751 4695] pits. Working did not finally cease at the former until 1949. The Darnley Fireclay, a short distance below the Calmy Limestone, was worked locally near Darnley. The fireclay was used to make high-grade sanitary ware. The Alum Shale occurring between the Hurlet Coal and the Hurlet Limestone was mined around Hurlet [513 611], from the Victoria Pit [521 506] and extensively under the South Brae of Campsie as a source of alum. The Lillie's Shale Coal near the top of the Lower Limestone Formation is a combined seam of coal and oil shale that was mined quite extensively in the Linwood area in the 19th century.

Hard rock

Two quartz-dolerite dykes were formerly quarried at Rashielee [465 709] for roadstone. Alkali dolerite sills have been quarried at a number of places in the Paisley–Johnstone area including Barr Hill, Kilbarchan, but the only quarry still in operation is High Craig [427 613] (Merritt and Elliot, 1984), where the products are used for roadstone and concrete aggregate. The reserves are considerable.

Lavas and associated intrusions

Lavas and associated vents and intrusions of the Clyde Plateau Volcanic Formation have been worked in several areas e.g. Craigengaun [525 769], Craigend [555 776] near Milngavie and Boylestone [492 598] at Barrhead. Currently, only Milton Hill [435 746], working a plug-like intrusion, and Dumbuckhill [420 747], working vent agglomerate and a dyke-like intrusion, are operating.

RESOURCES — DRIFT

Clays for refractory uses

A summary of clays of superficial origins in Central Scotland that are suitable for the manufacture of bricks is given by Elliot (1985). These resources are also suitable for pipe and tile manufacture. The clays in the Glasgow district are of several types including till, and late-Devensian glaciomarine and glaciolacustrine deposits. Flandrian estuarine, lacustrine and fluvial clays also are present but are of insufficient volumes to be significant resources.

Till is by far the most widespread Quaternary deposit in the district, both at surface and underlying younger deposits. Where till was worked for brickmaking, or canal bank puddleclay, the stones and boulders were usually handpicked. In brickmaking, bedrock mudstone was sometimes added. Till of the Wilderness Till Formation was quarried at Hamilton Hill [584 675] and Springburn Brickworks [605 672].

Late-Devensian glaciomarine clays of the Linwood and Paisley formations have been extensively worked for brick, tile, pottery and pipe manufacture in the Glasgow, Paisley and Rutherglen areas. The history of quarrying extends back several centuries with the last operation at Polmadie [604 624] only closing in the 1960s. When used for bricks, the clay was sometimes mixed with sand and ash. The clay in the Paisley area is known to make a good red facing-brick. It was formerly used at the Belvidere works in east Glasgow to manufacture strong sewer pipes. There are extensive resources remaining in the Linwood area which are still in greenfield locations. Both here and within the urbanised lower slopes of the lower Clyde valley, these clays are generally 15–30 m thick.

Late-Devensian glaciolacustrine clay of the Blane Water Formation has locally been worked for brick, tile and pipe manufacture just to the north of the district. These deposits extend southwards into Strath Blane, and are up to 20 m thick. However, they are located in an area of scenic value. In the south of Glasgow, at Williamwood [571 581], an older clay of the Bellshill Formation has been worked on a small scale.

Sand and gravel

A summary of resources of sand and gravel in the district is given in Cameron et al. (1977) and Browne (1977). The principal resources are confined to the valley floors and sides of the rivers Clyde and Kelvin, and the White Cart, Glazert, Endrick and Blane waters.

Moundy glacial sand and gravel is of little significance in the district and has only been exploited locally above the water table. The most extensive occurrences are near Strathblane [55 79] in the Broomhouse Formation, and near Gartness [49 86] in the Drumbeg Formation. The former are of Dimlington Stadial ice-contact glaciofluvial/glaciodeltaic origins, the latter are part of the local Loch Lomond Stadial terminal moraine and are of glaciolacustrine deltaic origin.

The most extensive sand and gravel deposits are mainly of Flandrian age and form the floodplains and terraces of the main rivers and streams. In the Clyde valley under Renfrew and Glasgow, the Quaternary succession, although dominated by glaciomarine clay, is commonly capped by about 6 m of sand, or sand and gravel. The deposit is commonly rich in coal debris and plant remains and is therefore of inferior quality. Historically, it has been exploited on a minor scale, including possibly for glass manufacture. The deposits of the White Cart, Endrick and Glazert waters may be generally coarser than those in the Clyde valley. They are also largely below the water table.

The most exploited deposits of sand and gravel in the district are those of the glaciofluvial/deltaic Cadder Formation. This formation is associated with the 'buried channel' of the River Kelvin. In general, the deposit is overlain by a variable thickness of till overburden (Wilderness Till Formation), probably the main reason why there are currently no active workings. The Cadder Formation is known to extend from Clydebank [50 71] to Cadder [61 72] and eastward into the Airdrie district. Now largely landfill, the main workings were at the Wilderness [600 725] and Cadder [606 721]. The former settling ponds at the Wilderness are currently exploited to provide 'fines' for cell construction at nearby active landfill sites.

Peat

Peat has been worked on a small scale on Cameron Muir as a domestic fuel and in historic times to burn concretionary limestone (cornstone) to produce lime. Although there are extensive deposits of hill peat in the Kilpatrick Hills and Campsie Fells, they are generally thin and also remote of access. The remnants of raised mosses in the Linwood and Paisley areas are as much as 6 m thick locally, but both Linwood [44 66] and Fulwood [445 690] mosses are partly covered by landfill. Together with Barochan Moss [430 685], these are now recognised as scarce ecological habitats.

RESOURCES — GROUNDWATER

Hydrogeology

Mean average annual rainfall is less than 1000 mm in Rutherglen and 1200 mm at Barrhead and Milngavie, but over 1800 mm on top of the Campsie Fells. Potential evaporation and transpiration by plants may amount to 350 to 400 mm in low-lying areas, so that 600 to 650 mm and 800 to 850 mm of effective rainfall is available at Rutherglen and Barrhead respectively. The greater part of the effective rainfall supports run-off and stream flow, but up to a third may become groundwater by infiltrating the soil and drift profile to reach the water table. This amount is equvalent to 200 to 280 mm/a day or enough to sustain a springflow of 9 l/s for each square kilometre of catchment. The widespread distribution of till within the district may, however, inhibit recharge in certain areas.

Groundwater has never been widely used in the district despite its favourable recharge potential. The reasons for this are the abundance of surface water supplies on the outcrop of the Clyde Plateau Volcanic Formation, the availability of piped water from Loch Katrine to Glasgow from 1859 onwards and the generally poor quality of groundwater within the Carboniferous sedimentary rocks.

Glasgow is the focal point for much of the groundwater discharge from the Central Coalfield (Robins, 1990) and therefore standing groundwater levels are near the surface throughout much of the city. The prevailing groundwater flow paths are from the east, northeast and south-east. The hydraulic conductivity of the Carboniferous rocks is of the order of 1 m/d in the more arenaceous horizons but may be only 10^{-2} m/d or less in the mudstones and coals. The groundwater flow potential of these strata is enhanced by the presence of faulting, fractures and abandoned mine workings.

Lavas and tuffs of the Clyde Plateau Volcanic Formation crop out to the north and south of Glasgow. These rocks form the Kilpatrick Hills and Campsie Fells to the north of the city, beyond which the Devonian sandstones straddle the valleys of the Endrick and Blane waters. Neither the lavas nor the sandstones offer any significant groundwater potential, although there are some isolated spring and borehole sources. Groundwater flow is predominantly local, from beneath high ground towards lower-lying areas, where discharge may occur as baseflow to the rivers.

In the district, there are no records of water boreholes which penetrate rocks of Lower Devonian age. There is, however, a group of nine exploratory water boreholes in these rocks some 3 km farther to the north-west near Balmaha (Bird et al., 1992). They all encountered Garvock Group channel-phase sandstones with some overbank mudstones. There are three groups of three boreholes: Balmaha car park [421 910], Milton of Buchanan [448 905] and Auchingyle Farm [430 907]. The transmissivity of the aquifer is 2 to 4 m^2/d and the storativity is 10^{-5}. Borehole specific capacity is, for the most part, less than 0.1 l/s/m and maximum sustainable yield from any one borehole is 3 l/s. Geophysical fluid logging identified fissure inflow to the boreholes and water quality which differed from one fissure to another. At shallow depths the water is Ca-HCO$_3$ type and deeper water is NaCl-SO$_4$(Cl) type. Sulphate levels are elevated due to the local presence of barite and the deeper more-saline waters may reflect the presence of a small residual element of Quaternary sea water.

The Upper Devonian and oldest Lower Carboniferous rocks are equally disappointing. In Fife, the Stratheden

Group contains up to 170 m of aeolian sandstone, the Knox Pulpit Formation, in which boreholes sustain abstraction of up to 40 l/s. In the west, the equivalent aeolian sandstones are interbedded with fluvial ones (Stockiemuir Sandstone Formation) and the prospects for groundwater development are far more modest. One successful borehole is recorded in the Upper Devonian in the district, at Glengoyne Distillery [526 825]. However, the sustainable yield is only 1 l/s and the borehole specific capacity is 0.2 l/s/m.

Some groundwater is present in cracks and joints in the Clyde Plateau Volcanic Formation. It is not generally feasible to drill boreholes to intersect the water-bearing fissures, although boreholes with sufficient capacity to support domestic premises do exist in parts of Renfrewshire. Wherever water-bearing fissures intersect the ground surface there may be spring flow; there are many springs and seepages. Water quality may be iron and manganese rich where oxygen has been depleted in older circulating waters. Equally, it may be weakly mineralised young water suitable for bottling, as is the case at Lennoxtown (a spring-field around 645 789) just to the east of the district.

Groundwater potential is poor in the overlying younger Carboniferous sedimentary rocks around Glasgow. At the base of the sequence, the sandstones of the Strathclyde and Inverclyde groups formerly sustained 2 and 6 l/s from adjacent boreholes in Paisley [457 628]. However, these yields are exceptional and are caused by the proximity of heavily faulted rock in the Paisley Ruck. Yields elsewhere in the sandstone rarely attained 1 l/s, although several of them overflowed when pumping ceased.

The majority of water boreholes in Glasgow penetrated the Lower Limestone, Limestone Coal or Upper Limestone formations. Sustainable abstraction rates were very small (less than 1 l/s) except where old mine workings were encountered. In 1985 a series of four boreholes were drilled at Govan for water supply to the Glasgow Garden Festival which was held the following year. The first borehole [5668 6475], 198 m deep, penetrated thin coaly mudstones, shale and sandstones belonging to the Upper Limestone Formation. The hole was tested at 9 l/s and the borehole specific capacity is 0.9 l/s/m. Three production boreholes followed [5645 6490; 5680 6487; 5730 6482], each with a production yield of only 1 l/s. The first borehole made contact with water-bearing fissures, yields from the other boreholes being entirely dependent on the intergranular permeability of the formation.

Water quality from the Carboniferous sedimentary rocks is poor, characteristically very hard and reducing with abundant iron and manganese in solution. Shallow mine workings encouraged dewatering of much of the near surface strata, allowing oxidation of pyrite to occur. Subsequent flooding rapidly took the newly formed hydrous iron oxides into solution. This process occurred throughout the 19th century; groundwater supplies with deteriorating quality were being abandoned from the 1860s onwards in favour of Loch Katrine mains supplies.

The response of potential users to the poor groundwater quality is illustrated by the brewers. Only three breweries, Hugh Baird & Co., Castle Brewery and J & R Tennant used groundwater, but all had to blend it with Loch Katrine water and none pursued its use longer than was economically necessary.

Groundwater in the drift deposits of the district has never been exploited. Permeable gravels are present around Bearsden and towards Balmore as well as along the valleys of the Endrick and Blane waters near Killearn and the River Clyde in Glasgow. The gravels in Glasgow offer a perched aquifer of limited resource potential, highly vulnerable to urban and industrial pollutants, not least leaking sewers. The gravels in the north, around Killearn, offer a better prospect, although vulnerable to agricultural pollutants; they are, however, only partly saturated and may be covered by alluvial silt and clay.

GEOLOGICAL HAZARDS

Coal workings

A particular problem in Glasgow is the presence of methane in solution in the groundwater. A 183 m-deep borehole into the Limestone Coal Formation to the north-east of the unversity [579 672] was described in 1939 by Dr A G MacGregor as follows:

'The bore gives off gas continuously, and to get rid of it a jet is kept burning. The flame is about 4 inches (100 mm) long, and yellowish. An iron pipe sticks up in a concrete trough in the courtyard. The diameter of the pipe is about 4 inches (100 mm); it is plugged at the top with two short bits of piping ($\frac{1}{2}$ to $\frac{3}{4}$ inches i.e. 12 to 18 mm in diameter) let into the plug. The gas burns at the top of these … tried putting cement over the hole but it bulged up by the gas pressure. The flame is protected from blowing out by a loose cylindrical collar'.

Other borehole records suggest that methane gas was a widespread problem, particularly in water from the Limestone Coal Formation.

Much of the coal beneath Glasgow has been removed by stoop-and-room working. Long abandoned and flooded, slow degradation of the pillars in these workings may lead to sudden ground failure. Cement injection and other packings are used to support both old and new buildings. These supports may divert groundwater to other routes so promoting chemical degradation and erosion of pillars elsewhere. The problem is illustrated by an inspection of 114 bore logs in the Hillhead area where only the Knightswood Gas Coal has been worked. The results of the study carried out by I H Forsyth are given in Table 8.

Table 8 Stability related to stoop-and-room workings.

State of Knightswood Gas Coal	No. of boreholes
Coal	35
Unspecified waste	28
Packed waste	24
Void	20
Collapsed waste	6
Faulted out	1

The Knightswood Gas Coal is generally not more than 80 cm thick. However, the presence of water-filled voids and collapsed wastes at shallow depths has serious implications for ground stability. The role of groundwater as an agent of degradation is unclear, but the opportunity exists for chemical interaction between water and rock, as well as physical erosion by flowing water.

Waste disposal sites

The hydraulic properties of the drift are of greater significance in terms of civil engineering, particularly with regard to landfill. Early landfill in the Glasgow area took place in the numerous sand and gravel and brick clay pits, and in building stone quarries in and around the city.

The search for waste disposal sites has now spread to and beyond the city limits. The Wilderness Landfill at Buchley [59 72] was formerly a sand and gravel pit. In the late 1970s it suffered leachate loss through the side walls into the River Kelvin, a problem resolved only by careful engineering as the landfill progressed. Another difficult site was at Balmuildy [574 716] on the flood plain of the River Kelvin. Again, sand and gravel lenses within the clayey drift allowed leachate access to the river. The adjacent landfill at Blackhill Brickworks [578 715] was engineered in the early 1980s to allow leachate access to the Upper Limestone Formation via the sands and gravels. Next door at Summerston [575 714], the philosophy had progressed by the mid-1980s to containment of leachate; an impermeable liner was used beneath the new landfill. All contemporary landfill development requires full containment engineering.

The hazards associated with the build-up of methane gas (CH_4) concentrations in landfill sites are now well appreciated (Campbell, 1988). Examples where domestic waste infill has subsequently been sealed by fine-grained cohesive soils, are numerous. Plans to extract gas commercially to heat greenhouses from landfill sites at Summerston [575 715], Wilderness [600 723] and Mavis valley [595 713] were in hand during 1993.

Contaminated land

There is a considerable residue of contaminated land in the Glasgow district. A great deal of reclamation and development has taken place on the less-contaminated sites, but several former gas works and chemical installations remain. Although infiltration of undesirable chemicals to the groundwater causes little threat to the few groundwater users, emergence as baseflow to rivers does cause a potential problem. However, in most cases it is almost impossible to trace individual sources of pollution due to the effect old mine workings have on prevailing groundwater flow patterns.

ELEVEN

Geophysical investigations

Geophysical data provide evidence for the interpretation of the deeper geological structure of the district. The Bouguer gravity and aeromagnetic surveys are the most comprehensive, but are complemented by a few seismic surveys. Gravity and aeromagnetic anomaly contour maps and the main anomalies referred to in the text are shown in Figure 29. The contours are based on values held in BGS databanks and are similar to those on the published 1:250 000 scale maps. The general anomaly pattern is simple. There are gravity lows, and low-frequency and low-amplitude magnetic anomalies over Upper Palaeozoic sedimentary rocks. Gravity highs with high-frequency and high-amplitude magnetic anomalies are associated with the Clyde Plateau Volcanic Formation. Many of the anomalies extend outside the district and feature in regional geophysical interpretations (e.g. Davidson et al., 1984; Evans et al., 1988). The major geophysical features are summarised in Figure 30. These can be compared with the mapped fault pattern (Figure 16).

GRAVITY

The gravity map of the district (Figure 29a) is based on surveys with a station distribution of about one per 2 km^2. The Bouguer anomaly values were referred to the 1973 National Gravity Reference Net, calculated against the 1967 International Gravity Formula, and reduced to sea level using a density of 2.70 Mg m^{-3} for the Bouguer correction.

In an early detailed survey McLintock and Phemister (1929) traced the morphology of the rockhead in the 'buried valley of the River Kelvin' near Drumry [5151 7101] at Clydebank. Large-scale work by Qureshi (1970) suggested that north-west of the Gartness Fault the Lower Devonian lay on Dalradian rocks at a depth of about 1.7 km. However, Qureshi used a relatively low density of 2.53 Mg m^{-3} (Table 9) for the Lower Devonian succession and probably considerably underestimated the depth to the base of the Devonian as a re-

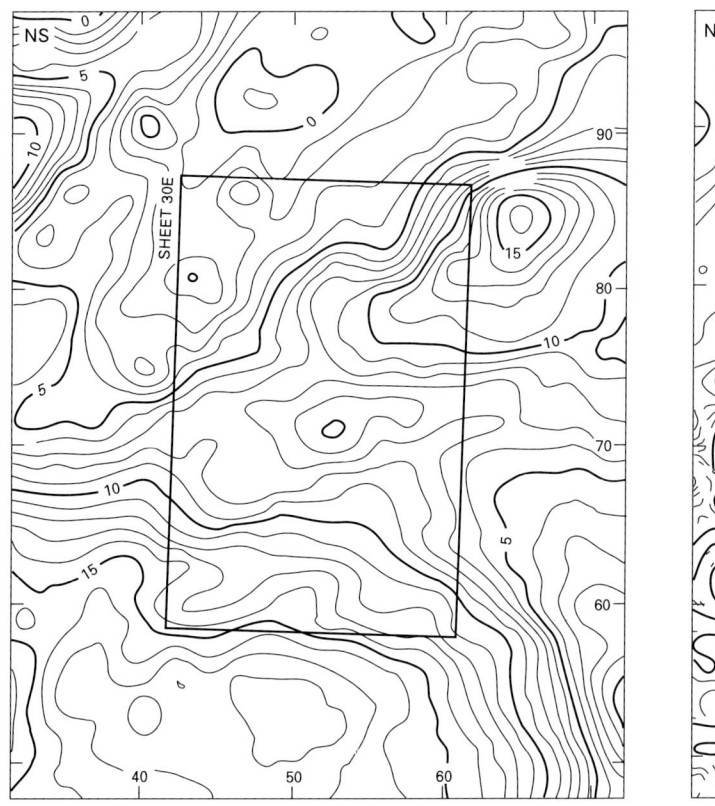

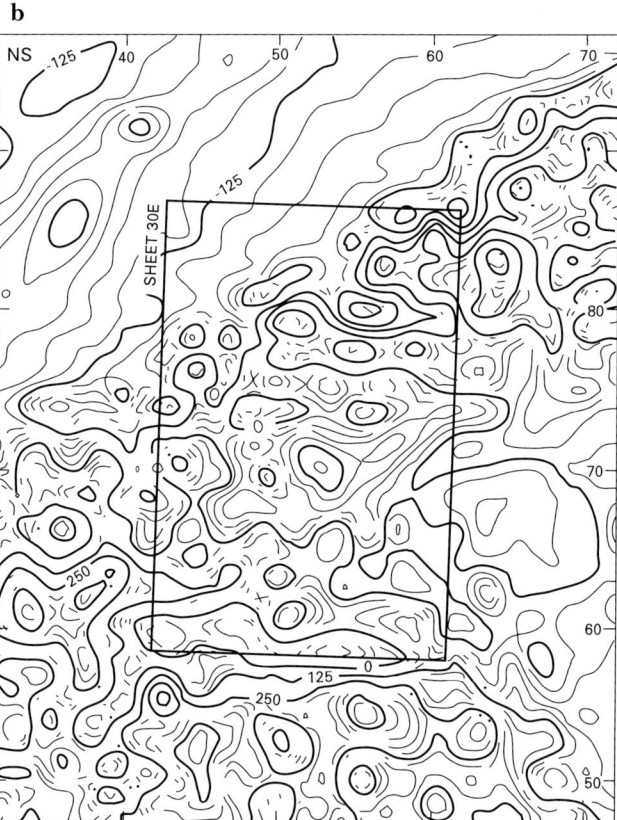

Figure 29 Geophysical maps of the Glasgow district and adjacent areas.
a. Bouguer gravity anomaly map with contours at 1 mGal interval. Density of 2.70 Mgm^{-3} used for Bouguer correction
b. Aeromagnetic total field anomaly map with contours generally at 25 nT interval, otherwise at 100 nT.

Figure 30 Main geophysical features of the Glasgow district and adjacent areas.

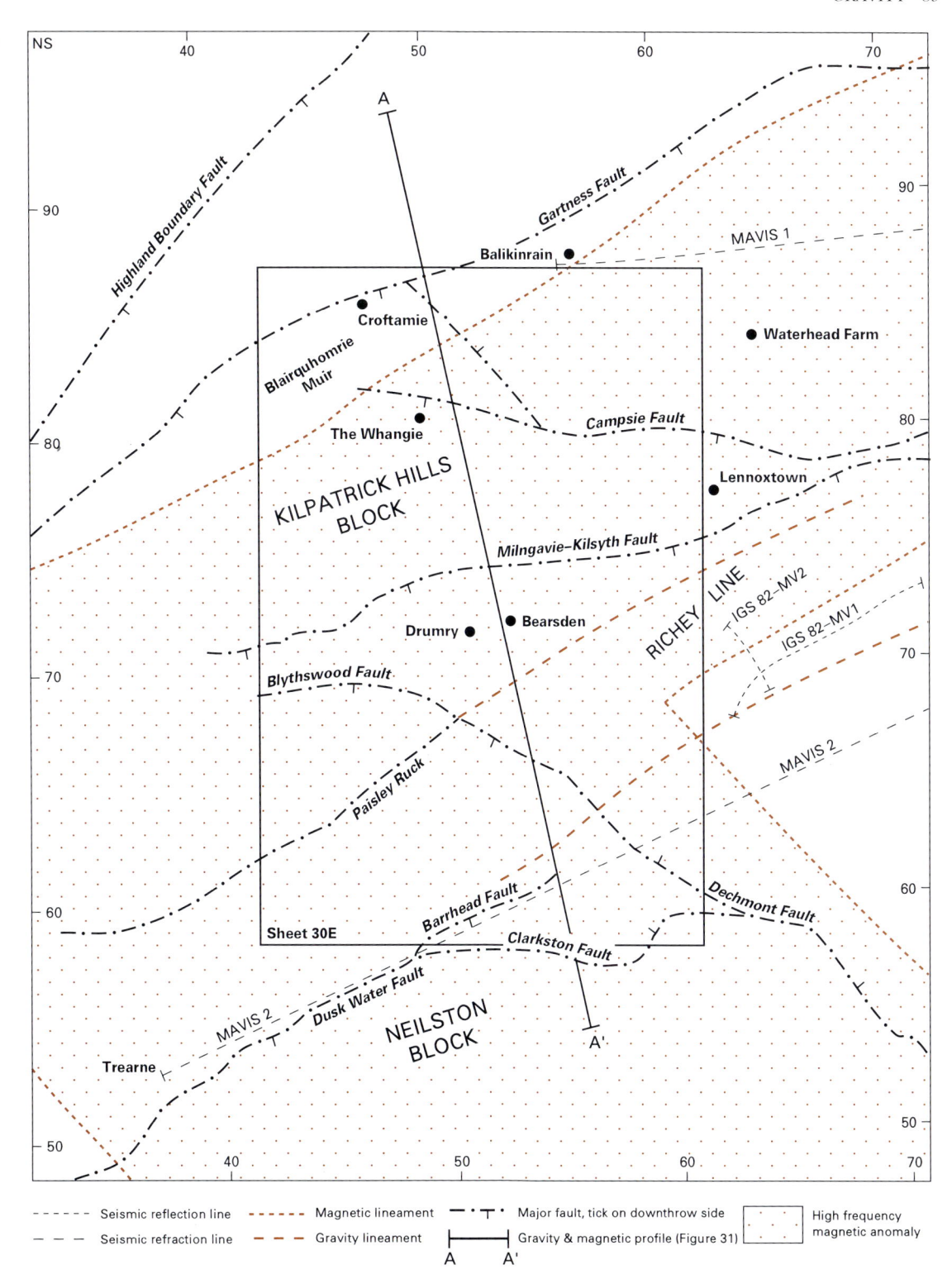

sult. A reinterpretation by Dentith et al. (1992) suggested that the Lower Devonian is only 200–400 m thick between the Highland Boundary and Gartness faults. They suggested that the Devonian may rest on the Ordovician Highland Border Complex, which possibly reaches depths of at least 2 km. They also believed the Highland Border Complex could extend for a considerable distance into the Midland Valley, overlying a non-Dalradian crystalline basement. The Croftamie high was attributed to a 900 m-thick igneous body at the base of the Lower Devonian north-west of the Gartness Fault. The gravity high is not associated with a magnetic anomaly. This suggests that the inferred causative igneous body may be dioritic which would be in agreement with its

Table 9 Representative physical properties for saturated rocks at ground level in the Glasgow district, based on sample testing.

Formation	Saturated density Mg m^{-3}	Magnetic susceptibility 10^{-3} (SI)	Remanent intensity A m^{-1}	P-wave velocity km s^{-1}
Quaternary (clays and sands)	1.72–2.24	— N	— N	—
CARBONIFEROUS				
Westphalian	2.50	— N	— N	2.0–4.5
Namurian	2.55	— N	— N	3.2–3.6
Dinantian	2.55	— N	— N	2.0–3.8
Clyde Plateau Volcanic Formation, and other Dinantian lavas (largely alkalic)	2.74 (2.47–2.98)	30–100	1–10	2.8–4.8
DEVONIAN				
Upper Devonian	2.40–2.55	— N	— N	2.0–4.0
Lower Devonian sedimentary rocks	2.53–2.63	— N	— N	3.0–4.0
Lower Devonian andesitic lavas	2.66–2.72	1–10	0.1–1.0	4.0–6.0
LOWER PALAEOZOIC				
Southern Uplands greywackes	2.72	N	N	3.7–4.3
Highland Border Complex	2.68	—	—	—
Dalradian metamorphic rocks	2.72	N	N	3.9
PRECAMBRIAN				
Lewisian granulites amphibole	2.74	2.6	N	5.64
Lewisian granulites pyroxene	2.86	26	N	6.34
IGNEOUS INTRUSIONS				
Late Carboniferous–Early Permian quartz-dolerite	2.80	1–10	1	5.6
Carboniferous and Permian alkali dolerite (olivine-dolerite, teschenite, kylite, monchiquite)	2.85	35	1	6.0
Upper Palaeozoic basic (gabbroic)	2.82	30	—	—
Caledonian diorite	2.7–2.8	20	—	5.6

Sedimentary rocks unless otherwise stated. — = not determined; N = probably negligible.

calculated density. The intrusion would be of Devonian age.

The Clyde Plateau Volcanic Formation in the Campsie Fells and Kilpatrick, Renfrewshire and Beith–Barrhead hills is associated with gravity and magnetic highs (Figure 29). Interpretations of these anomalies suggest they cannot be caused just by the inferred thickness (400–800 m) of lavas present (Cotton, 1968; Evans et al., 1988). It has been suggested that the anomalies are probably partly due to related underlying 1.2–1.8 km-thick basic (?gabbroic) intrusions.

Cotton (1968) interpreted two detailed traverses from the Kilpatrick Hills Block to the Beith–Barrhead and Neilston blocks. He estimated the Milngavie–Kilsyth Fault, which bounds the Kilpatrick Hills Block in the south, to have a southerly downthrow of about 200 m. The fault is marked by a zone of decreasing gravity into a low, here called the Bearsden low. To the west of this low, Cotton calculated the Blythswood Fault to throw down the top of the Clyde Plateau Volcanic Formation about 330 m to the south, whereas surface evidence indicates a throw of 150 m.

The Bearsden low is bounded to the south by a southerly increasing gradient, the south-west part of which coincides with the Paisley Ruck. Cotton estimated the ruck to have a north-westerly downthrow of about 140 m and that the lavas thickened south-east across the ruck by about 200 m. The maximum thickness of lavas was interpreted to be 1.2 km on the downthrown (north) side of the Clarkston Fault (which joins the Castlemilk Fault), where the throw was 900 m. The gradient marking the Paisley Ruck extends north-east of the Blythswood Fault but is not associated with any structure known at the surface. However, the ruck may continue at depth to the Campsie Fault near Lennoxtown [686 764]. This possible extension would downthrow the Clyde Plateau Volcanic Formation to the north-west at the south-east margin of the Bearsden low. Cotton (1968) showed a depression in the top and bottom surfaces of the Clyde Plateau Volcanic Formation under the Bearsden low, but no faulting on its south margin. In the centre of this basin, the top of the lavas appeared to be 630 m deep.

The extension of the gradient flanks the north-west side of a gravity high over the Riggin Anticline which occurs in the Airdrie district (Figure 29). A subparallel gradient trends north-east from the Dusk Water–Barrhead Fault to near Cumbernauld [746 728]. The Dusk Water–Barrhead Fault throws down Carboniferous sedimentary rocks to the south-east and thus the associated gradient may be due to a continuation of this fault zone throwing down the underlying Clyde Plateau Volcanic Formation in the same direction. This gradient is broadly coincident with the Richey Line (Read, 1988, fig. 16.1). The Riggin Anticline does not extend to the Dusk Water–Barrhead Fault but, east of the Dechmont Fault, there is an anticline between the continuations of the Dusk Water–Barrhead Fault (the Richey Line) and the Paisley Ruck.

MAGNETIC

The total field aeromagnetic map (Figure 29b) is based on data recorded at 305 m above ground level. Flights were made on a grid, east–west survey lines 2 km apart and north–south ties 10 km apart. The observed data were processed for the removal of a first-order regional field. The regional magnetic pattern suggests that the crystalline basement of the Midland Valley is of Lewisian type, possibly pyroxene, overlain by amphibole-granulites (Powell, 1970; 1978). Such rocks have been reported from vents in the eastern Midland Valley (Upton et al., 1980).

A dipolar anomaly is associated with the boundary between the Clyde Plateau Volcanic Formation and the Devonian sedimentary rocks. The anomaly has a trough to the north and peak to the south. This shows that the lavas in bulk have normal total magnetisation (Francis et al., 1970), which is probably of largely induced origin. Measurements on samples indicate that the natural remanent magnetisation (NRM) of the lavas (Palmer et al., 1985) is largely reversed but that the directions are too scattered and their intensities too weak to contribute significantly to the total magnetisation.

Magnetic anomalies over the Carboniferous sedimentary rocks are like a smoothed version of those over the Clyde Plateau Volcanic Formation. This indicates that the latter, and/or Devonian lavas, lie at depth. The field is smoothest over the Devonian sedimentary rocks, indicating that they are not intruded by large basic igneous bodies. The many Carboniferous vents within the Devonian do not create aeromagnetic anomalies, probably because the vents are small and the agglomerates are only weakly magnetic.

A magnetic low is associated with the Bearsden gravity low. It is bounded by a magnetic high to the south of the postulated extension of the Paisley Ruck (Figure 29), supporting the gravity based inference that the Paisley Ruck extends north-east of the Blythswood Fault. The magnetic map does not clearly show the inferred extension of the Dusk Water–Barrhead Fault, perhaps because the survey excluded the centre of Glasgow. The character of the anomalies here called the Waterside and Cathcart highs is consistent with the sources being concealed Carboniferous igneous bodies, probably the Clyde Plateau Volcanic Formation.

Ground magnetic traverses across the approximately 30 m-wide east–west Permo-Carboniferous quartz-dolerite dykes show dipolar anomalies, with amplitudes of 1000–5000 nT (Powell, 1963; Cotton, 1968; Maxwell, 1971). The troughs were generally south of the peaks, indicating reversed total magnetisation due to closely grouped NRM directions (Powell, 1963; Cotton, 1968).

Maxwell (1971) found strong magnetic anomalies over the bedrock depressions associated with the valley of the River Kelvin. The anomalies were thought to be due to Quaternary magnetic sediments since the Devonian and Carboniferous sedimentary rocks in the district are essentially non-magnetic.

SEISMIC

The acoustic impedance of the Clyde Plateau Volcanic Formation is generally much higher than that of sedimentary rocks so that its boundaries have high reflection coefficients, preventing much energy from penetrating to greater depths. The usual velocity inversion at its base inhibits detection of that surface in conventional refraction exploration.

Reflection

There are no reflection surveys in the district, but surveys in adjacent areas include the IGS82 MV1 and MV2 Vibroseis lines recorded to 6 s two-way time (Penn et al., 1984, figs 3.2 and 4.1). Such surveys generally show coherent reflectors only to 1–2 s two-way time, equivalent to depths of 1.5–3.0 km and seldom below the base of the Clyde Plateau Volcanic Formation (Hall, 1971; 1974; Evans et al., 1988). Penn et al. (1984) found the lavas to be approximately 600 m thick, with their top about 1 km deep. The deepest reflector, at about 2 km, was interpreted as the top of the Lower Devonian.

Andrew (1978) reported an experimental deep reflection survey near Kippen [5807 9130] where depths for the bases of the Devonian and Lower Palaeozoic rocks were estimated as 4.4 km and 8.6 km respectively. The 4.4–8.6 km interval had a velocity of 6.4 km s^{-1} (Table 9), suggesting that it is Lewisian crystalline basement rather than Lower Palaeozoic rocks. The 4.4 km-deep reflector could then be interpreted as the base of the Devonian, as supposed by Andrew (1978), or the base of the Lower Palaeozoic.

Refraction and wide-angle reflection

The early LISPB and LOWNET refraction surveys were of low resolution, but provided deep penetration of 20–40 km (Bamford et al., 1978). Although centred east of the district, their results are relevant to the whole of the Midland Valley. LISPB revealed a 5.8 km s^{-1} refractor ('a$_0$'), at a depth of about 3 km, thought to be the top of the Lower Palaeozoic (Bamford et al., 1978; Barton, 1992). A 6.4 km s^{-1} refractor ('a$_1$'), about 7–8 km deep, was interpreted as the top of crystalline basement of Lewisian type.

These surveys were followed by the shorter MAVIS and quarry blast surveys, of which only MAVIS 2 is located in the district (Figure 30). The MAVIS surveys detailed refractors to depths of about 12 km (Conway et al., 1987; Dentith and Hall, 1989). The quarry blast surveys had the shortest lines, giving detailed information to depths of 3–6 km (Davidson et al., 1984; Dentith and Hall, 1990). These surveys have been interpreted in terms of four main refracting layers, two of which ('a$_0$' and 'a$_1$') are now differently identified from the original LISPB interpretation. This implies a significantly thinner Lower Palaeozoic succession and a shallower crystalline basement. The four layers are as follows:

1) 0.5–3.0 km thick Carboniferous and Upper Devonian strata (velocities in the range 3–5 km s^{-1})
2) 1–5 km thick Lower Devonian and Lower Palaeozoic strata (4.8–5.4 km s^{-1})
3) LISPB layer 'a$_0$', 3–4 km thick, top at 3–6 km, gneissose crystalline basement (5.8–6.2 km s^{-1})
4) LISPB layer 'a$_1$', top 7–9 km deep; an intra-basement refractor, possibly due to a downward increase in metamorphic grade from amphibole to pyroxene granulite facies, (6.4 km s^{-1})

Interpretations of the MAVIS and quarry blast surveys by Dentith and Hall (1989; 1990) suggested *décollement* at the top of the basement and, locally, in layer 1. Faults in layer 1 may be listric, soling out into the possible *décollement* at its base (Dentith and Hall, 1990). In the Glasgow district the tops of layers 2, 3 and 4 on MAVIS 2 are at depths of about 2.5, 4 and 7.5 km respectively. The suggested nature of the basement is in accord with that inferred from magnetic indications.

Resistivity

Detailed resistivity and seismic refraction surveys have been used in the area for assessing rock masses for quarrying (Wattananikorn, 1978).

GEOTHERMAL

Basic geothermal information for the district is in catalogues, of which the latest is Rollin (1987a). The geothermal potential of the Midland Valley is summarised in Browne et al. (1985; 1987), Rollin (1987b) and Evans et al. (1988). The average geothermal gradient is 22.5°C km^{-1}, and the heat flow 54.5 mW m^{-2}, but higher values occur in the district (Browne et al., 1987).

Only two prospects with geothermal potential have been recognised above the basement. These are:

1) sandstones at the top of the Devonian, mainly the Knox Pulpit Formation (equal in part to the Stockiemuir Sandstone Formation)
2) sandstones of the Passage Formation

In the district, both of these are deepest in the southeast, where the latter may reach only 600 m in depth and an estimated temperature of 20°C, but the former may exceed 1.5 km in depth and a temperature of 40°C (Browne et al., 1985; 1987).

PHYSICAL PROPERTIES

The physical properties of the main rock units are summarised in Table 9. Generally, density and velocity increase significantly with depth. There are no local exposures of crystalline basement to allow determination of its properties.

Above the basement the first large density and velocity contrast may be between Lower Palaeozoic and Lower Devonian rocks. However, the lithologies of large parts of the Silurian sequences in the Lesmahagow and other inliers are more akin to the Lower Devonian sedimentary rocks of the Midland Valley than to the Silurian greywackes of the Southern Uplands (Paterson et al., in press). The average density of the Lower Devonian succession, taking into account a contribution from lavas, could approach that of the Highland Border Complex. Thus, in the Midland Valley the Lower Palaeozoic may have a density of 2.68 Mg m^{-3}, with a corresponding velocity, and hence be indistinguishable from the Lower Devonian.

The next density and velocity contrast is between Lower Devonian rocks and overlying Upper Devonian and Lower Carboniferous sedimentary rocks. The third is between the latter two and the Clyde Plateau Volcanic Formation. The last contrast is between these lavas and overlying Carboniferous sedimentary rocks. Average values for the Devonian and Carboniferous successions vary and depend upon the poorly known ratios of igneous to sedimentary rock and sandstone to mudstone.

The NRM of Devonian rocks in Scotland is complicated, and its interpretation controversial (e.g. Torsvik, 1985; Storetvedt et al., 1991; Walderhaug et al., 1991; references therein). Apparently, these rocks record multiple primary (Devonian) and secondary (post-Devonian) NRM directions. Possibly the commonest primary direction is reversed, with a declination of about 220°, and an inclination of about +40°. However, there is probably another primary component directed reciprocally to this, providing evidence of polarity reversal. The intensity of all these components relative to each other and to the induced intensity is uncertain. Probably, the remanent magnetisations largely self-cancel, because of the scatter in their directions, making for only a small net (total) magnetisation due to the dominant induced magnetisation. The Clyde Plateau Volcanic Formation shows the same scattered NRM and has a strong total magnetisation due to high susceptibilities.

Carboniferous and Permian igneous rocks generally have reversed NRM, with a declination of about 180° and inclinations between +30° and -20° (Torsvik et al., 1989; references therein). Detailed study of the Lenzie–Torphichen Permo-Carboniferous quartz-dolerite dyke shows significant changes in the ratio of remanent to induced magnetisation along strike, creating variation in the form of the magnetic anomalies (Powell, 1963).

Overall, the Devonian, Carboniferous and Permo-Carboniferous NRM directions suggest movement of the Midland Valley terrane during Carboniferous times from low southerly latitudes to north of the equator in the Permian.

INTERPRETATION PROFILE

A profile has been constructed across the district (Figure 31) using the relationships between the gravity and magnetic fields and the surface geology to interpret the deeper structure. Geological control for the profile was provided by the two horizontal sections on the Glasgow Solid Sheet margin. The observed geophysical values were calculated from observations, held in a BGS digital databank, made up to 2 km away on both sides of the profile. A density of 2.70 Mg m^{-3} was used for the calculation of the Bouguer anomalies and the interpretation made using the GRAVMAG program of Pedley (1991), which assumes the bodies have finite strike extent. The profile was oriented to cross the strike of particularly interesting geological bodies.

Lower Palaeozoic rocks were assumed to form a background unit which is present throughout the district extending to a depth of at least 4.0 km. On the northwest projected line of the profile they crop out as the Highland Border Complex. The Lower Palaeozoic rocks possibly occur at a very shallow depth immediately southeast of the Highland Boundary Fault, and may extend a considerable distance into the Midland Valley (Dentith et al., 1992). To the south-east of the district, Lower Palaeozoic rocks crop out in the Lesmahagow and other Silurian inliers, where they probably have densities similar to those of the Highland Border Complex. Therefore, a density of 2.68 Mg m^{-3} was adopted for the Lower Palaeozoic of the Midland Valley. This is significantly different to the values used in previous interpretations, which generally assumed that the Lower Palaeozoic of the Midland Valley was similar in lithology and density to the greywackes of the Southern Uplands. Over exposed and near-surface Lower Palaeozoic rocks, the observed gravity field has values of about 1.0 and 19.0 mGal at distances along the profile of, respectively, -3.5 and +63 km. This variation defines a sloping linear 'regional' field, considered to be due to rocks below the Upper Palaeozoic. This was removed from the observed data to leave a 'residual' gravity field attributable to, and interpreted in terms of, rocks above the Lower Palaeozoic. This regional field is similar to those commonly used in the Midland Valley (e.g. Dentith et al., 1992).

A residual magnetic profile was produced by removing from the observed field the regional field that Hall and Dagley (1970) produced by filtering. Their regional field is attributed to deep and broad bodies that produce magnetic anomalies with a wavelength greater than 13 km and, like the gravity regional field, shows a NE–SW high along the centre of the Midland Valley. The resulting residual magnetic field is particularly appropriate for studying the effect of faulting in those igneous rocks that are shallow and thin. Some rocks of the Highland Border Complex are significantly magnetic compared to the Lower Palaeozoic of the Lesmahagow Inlier but, along the profile, they should be sufficiently deep and broadly distributed for their effect to be removed in the regional magnetic field.

The new gravity and magnetic interpretation across the district (Figure 30) is in general agreement with the large body of previous geophysical interpretations made in the area and does not contradict the known geology.

At the north end of the profile, the thickness of Devonian strata inferred is much greater than that suggested by Dentith et al. (1992), but is supported by the Kippen seismic results. To avoid a slight mismatch with the residual magnetic field it was necessary to include lavas within the Devonian succession; this is again supported by the Kippen results. The floor of the Devonian basin appears to dip steeply towards the axis of the Strathmore Syncline which may indicate that its base is downfaulted at depth, or forms a growth fold. Only the eastern margin of the Croftamie anomaly is intersected by the profile, so that it cannot be modelled accurately.

South of the Campsie Fault, the Clyde Plateau Volcanic Formation is inferred to occur everywhere, but at varying depths. These lavas are so magnetic that the model is fairly insensitive to deeper magnetic rocks. In the Kilpatrick and Beith–Barrhead hills their disposition, and that of underlying rocks, accords with previous interpretations.

Between the Kilpatrick and Beith–Barrhead hills the calculated magnetic field is about 100 nT greater than the observed. Clearly, there are magnetic rocks in this area whose effect has not been accounted for in the regional field, and which could be modelled. No bodies above the Lower Palaeozoic could be produced to account for the discrepancy. Deeper igneous intrusions, additional to those modelled, could be one possible cause, Lewisian granulites another.

In the same area there are significant differences between previous interpretations and this one. The thickness (about 300 m) of Carboniferous lavas is a little less than in earlier gravity and seismic interpretations but thicker lavas produced a large mismatch to the observed magnetic field. For the same reason the lavas are here inferred to be slightly deeper (about 1500 m) than in Cotton's (1968) gravity interpretation.

The magnetic trough on the north-west side of the Neilston Block is very similar to that on the north-west side of the Kilpatrick Block. However, it could not be modelled the same way. This is due to downfaulting of Carboniferous lavas by the Clarkston Fault. Carboniferous alkali dolerite sills crop out nearby, and the anomaly has been modelled as a sill at a shallow depth. However, it was necessary to give the sill remanence properties that are unusual in both their high strength and their normal direction.

94 ELEVEN GEOPHYSICAL INVESTIGATIONS

Figure 31 Gravity and magnetic profile across the Glasgow district and surrounding areas, interpreted by 2.5 dimensional modelling.

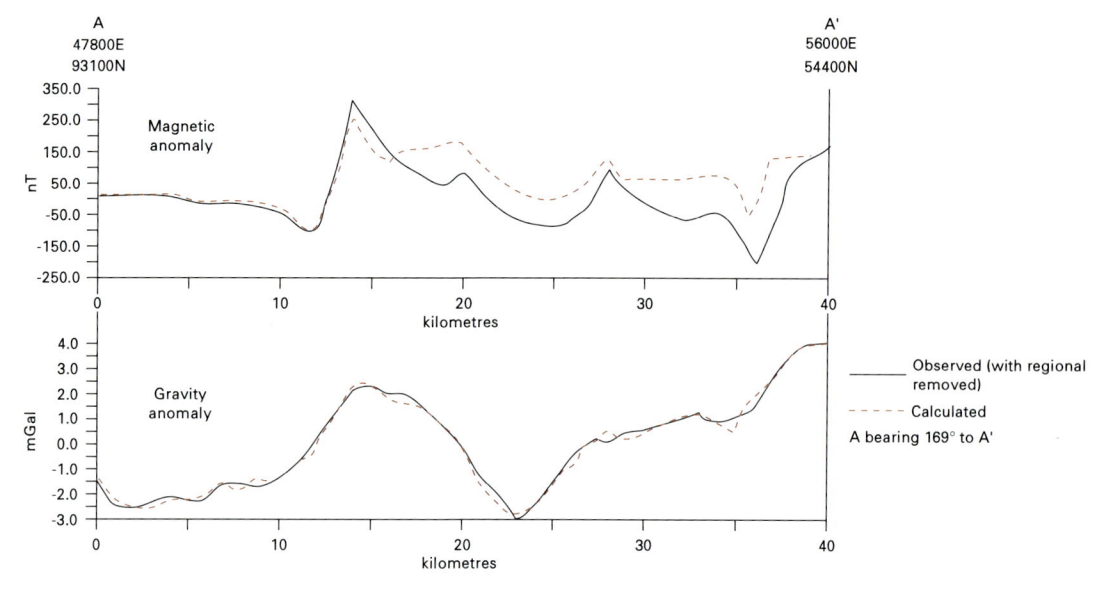

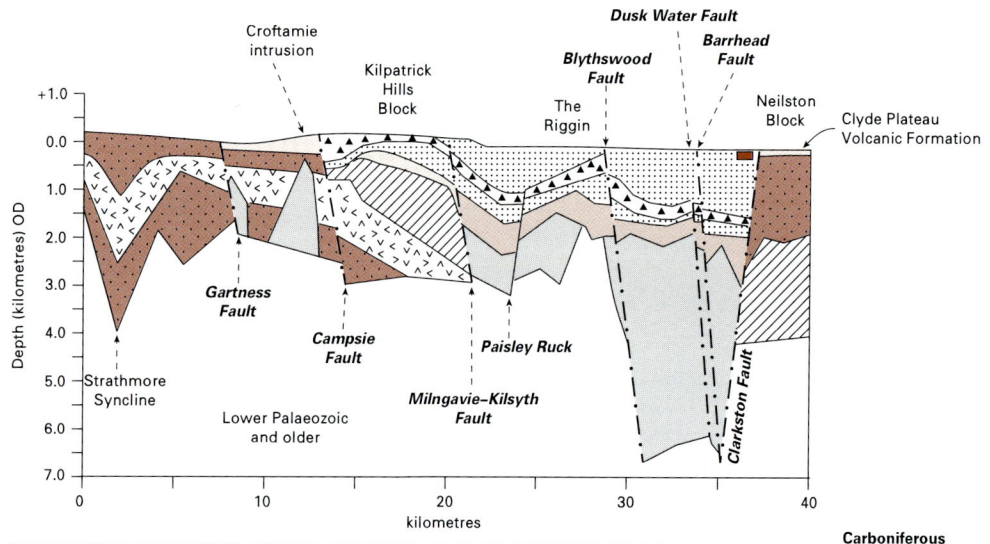

	Density Mg m⁻³	Magnetisation			
		Induced intensity A m⁻¹	Remanent (NRM)		
			Intensity A m⁻¹	Dec. degrees	Inc. degrees
CARBONIFEROUS					
Westphalian ⎱ sedimentary					
Namurian ⎰ rocks	2.25	0	0	—	—
Dinantian					
Clyde Plateau Volcanic Formation	2.74	1.925	1	180	15
DEVONIAN					
Undifferentiated	2.63	0	0	—	—
Upper	2.48	0	0	—	—
Lower ⎧ sedimentary rocks	2.63	0	0	—	—
⎩ andesitic lavas	2.68	0.166	0.1	220	40
LOWER PALAEOZOIC	2.68	0	0	—	—
IGNEOUS INTRUSIONS					
Carboniferous alkali dolerite	2.85	0.385	4	0	-15
Upper Palaeozoic basic (gabbroic)	2.82	1.155	0.25	200	28
Caledonian diorite	2.80	0.385	0.1	210	34

There is no information available on the possible relative thicknesses of the three Devonian units of Table 9 between the Kilpatrick and Beith–Barrhead hills. Therefore, the physical properties ascribed to the undifferentiated Devonian are biased towards those of the Lower Devonian sedimentary rocks. The total thickness of the Carboniferous and Devonian is consistent with seismic interpretations. The profile shows thickness variations in the Palaeozoic strata between the Kilpatrick and Beith–Barrhead hills, but modelling was not sensitive to them; Cotton's (1968) inferences about thickness variations in the Carboniferous lavas were not substantiated. However, the modelling was sufficiently sensitive to suggest that both the Paisley Ruck and the Dusk Water Fault continue north-east at depth.

Between the Kilpatrick and Beith–Barrhead hills the modelling suggests the presence of a substantial thickness of dense but moderately magnetic rocks. The interpreted thickness of sills (Carboniferous to Permian quartz- or alkali dolerite) necessary to account for the gravity anomaly far exceeded those currently known and, in addition, quartz-dolerite sills are not expected in this area and alkali dolerite sills would be too magnetic. Extensions of the basic intrusions underlying the Kilpatrick and Beith–Barrhead hills would also be too magnetic. The dense, moderately magnetic rocks are therefore interpreted either as Caledonian or Carboniferous diorites underlying the Upper Palaeozoic. These igneous bodies could even be parts of one intrusion, with those occurring under the Kilpatrick and Beith–Barrhead hills representing the basic marginal phases of the intrusion and the bodies between the blocks less basic phases towards the centre. What are shown as Lower Palaeozoic rocks at the base of the Upper Palaeozoic at 28 km along the profile, might then actually be a central, weakly magnetic granitic phase to the intrusion barely distinguishable in density from the Lower Palaeozoic. An alternative source of dense and moderately magnetic rocks here could be amphibolitised Lewisian granulites in a horst. However, to form a plausible structural model, they would have to be quite differently positioned to the 'Caledonian diorite'.

REFERENCES

Most of the references listed below are held in the libraries of the British Geological Survey at Murchison House, Edinburgh and at Keyworth, Nottingham. Copies of the references can be purchased from the Keyworth office subject to the current copyright legislation.

ANDREW, E M. 1978. Geothermal exploration in the Midland Valley of Scotland: some seismic reflection tests. *Report of the Applied Geophysics Unit, Institute of Geological Sciences*, No. 64.

ANDREWS, J E, TURNER, M S, NABI, G, and SPIRO, B. 1991. The anatomy of an early Dinantian terraced floodplain: palaeo-environment and early diagenesis. *Sedimentology*, Vol. 38, 271–287.

ARMSTRONG, M, PATERSON, I B, and BROWNE, M A E. 1985. Geology of the Perth and Dundee district. *Memoir of the British Geological Survey*, Sheets 48W, 48E, 49 (Scotland).

ASPEN, P. 1974. Fish and trace fossils from the Upper Old Red Sandstone of Dunbartonshire. *Proceedings of the Geological Society of Glasgow*, Session 113, 4–7.

ASPEN, P, and JARDINE, W G. 1968. A temporary exposure of Quaternary deposits at Renfrew, near Glasgow. *Proceedings of the Geological Society of Glasgow*, Session 109, 35–37.

BAMFORD, D, NUNN, K, PRODEHL, C, and JACOB, B. 1978. LISPB-IV. Crustal structure of Northern Britain. *Geophysical Journal of the Royal Astronomical Society*, Vol. 54, 43–60.

BARTON, P J. 1992. LISPB revisited: a new look under the Caledonides of northern Britain. *Geophysical Journal International*, Vol. 110, 371–391.

BELL, D. 1871. On the aspects of Clydesdale during the Glacial Period. *Transactions of the Geological Society of Glasgow*, Vol. 4, 63–69.

BELL, D. 1874. Notes on the glaciation of the west of Scotland, with reference to some recently observed instances of cross-striation. *Transactions of the Geological Society of Glasgow*, Vol. 4, 300–310.

BELT, E S, FRESHNEY, E C, and READ, W A. 1967. Sedimentology of Carboniferous cementstone facies, British Isles and eastern Canada. *Journal of Geology*, Vol. 75, 711–721.

BENNIE, J. 1866. On the surface geology of Glasgow. *Transactions of the Geological Society of Glasgow*, Vol. 2, 100–115.

BENNIE, J. 1868. On the surface geology of the district around Glasgow, as indicated by the journals of certain bores. *Transactions of the Geological Society of Glasgow*, Vol. 3, 133–148.

BIRD, M J, BUCKLEY, D K, CHENEY, C S, HALL, I H S, and ROBINS, N S. 1992. Groundwater investigations within the Devonian rocks near Balmaha, Central Region. *British Geological Survey Technical Report*, WD/92/2.

BLUCK, B J. 1978. Sedimentation in a late orogenic basin: the Old Red Sandstone of the Midland Valley of Scotland. 249–278 *in* Crustal evolution in northwestern Britain and adjacent regions. BOWES, D R, and LEAKE, B E (editors). *Special Issue of the Geological Journal*, No. 10.

BLUCK, B J. 1980. Evolution of a strike-slip controlled basin, Upper Red Sandstone, Scotland. 63–78 *in* Sedimentation in oblique-slip mobile zones. BALLANCE, B F, and READING, H G (editors). *Association of Sedimentology, Special Publication*, No. 1.

BOWEN, D Q. 1978. *Quaternary geology*. (Oxford: Pergamon.)

BOYD, W E. 1982. The stratigraphy and chronology of Late Quaternary raised coastal deposits in Renfrewshire and Ayrshire, western Scotland. Unpublished PhD thesis, University of Glasgow.

BOYD, W E. 1986. Vegetation history at Linwood Moss, Renfrewshire, central Scotland. *Journal of Biogeography*, Vol. 13, 207–223.

BRADY, G S, CROSSKEY, H W, and ROBERTSON, D. 1874. *A monograph of the Post-Tertiary Entomostraca of Scotland*. (London: Palaeontographical Society.)

BRAND, P J. 1977. The fauna and distribution of the Queenslie Marine Band (Westphalian) in Scotland. *Report of the Institute of Geological Sciences*, No.77/18.

BRITISH GEOLOGICAL SURVEY. 1994. Glasgow. Scotland Sheet 30E. Drift. 1:50 000. (Southampton: Ordnance Survey for British Geological Survey.)

BROWNE, M A E. 1977. Sand and gravel resources of the Central Region, Scotland. *Report of the Institute of Geological Sciences*, No. 77/9.

BROWNE, M A E. 1985. Comments on the Quaternary deposits and landforms of Scotland and the neighbouring shelves: a review (by D G Sutherland). *Quaternary Science Reviews*, Vol. 4, Pt 2, i-iii.

BROWNE, M A E. 1986. The classification of the Lower Carboniferous in Fife and Lothian. *Scottish Journal of Geology*, Vol. 22, 422–425.

BROWNE, M A E, and GRAHAM, D K. 1981. Glaciomarine deposits of the Loch Lomond stade glacier in the Vale of Leven between Dumbarton and Balloch, West-Central Scotland. *Quaternary Newsletter*, Vol. 34, 1–7.

BROWNE, M A E, GRAHAM, D K, and GREGORY, D M. 1984. Quaternary estuarine sediments in the Grangemouth area, Scotland. *Report of the British Geological Survey*, Vol. 16, No. 3.

BROWNE, M A E, HARGREAVES, R L, and SMITH, I F. 1985. *The Upper Palaeozoic basins of the Midland Valley of Scotland. Investigation of the Geothermal Potential of the UK.* (Keyworth, Nottingham: British Geological Survey.)

BROWNE, M A E, HARKNESS, D D, PEACOCK, J D, and WARD, R G. 1977. The date of deglaciation of the Paisley–Renfrew area. *Scottish Journal of Geology*, Vol. 13, 301–303.

BROWNE, M A E, and McMILLAN, A A. 1984. Shoreline inheritance and coastal history in the Firth of Clyde. *Scottish Journal of Geology*, Vol. 20, 119–120.

BROWNE, M A E, and McMILLAN, A A. 1989a. Quaternary geology of the Clyde Valley. *British Geological Survey Research Report*, SA/89/1.

BROWNE, M A E, and McMILLAN, A A. 1989b. Geology for Land Use Planning: drift deposits of the Clyde valley. Volume 1 Planning Report; Volume 2 Details of procedures and technical details; Volume 3 Thematic maps. *British Geological Survey Technical Report*, WA/89/78.

BROWNE, M A E, MCMILLAN, A A and GRAHAM, D K. 1983a. A late-Devensian marine and non-marine sequence near Dumbarton, Strathclyde. *Scottish Journal of Geology*, Vol. 19, 229–234.

BROWNE, M A E, MCMILLAN, A A and HALL, I H S. 1983b. Blocks of marine clay in till near Helensburgh, Strathclyde. *Scottish Journal of Geology*, Vol. 19, 321–325.

BROWNE, M A E, and eight others. 1987. *The Upper Devonian and Carboniferous sandstones of the Midland Valley of Scotland. Investigation of the geothermal potential of the UK.* (Keyworth, Nottingham: British Geological Survey.)

BROWNE, M A E, and five others. 1996. A review of the lithostratigraphy of the Carboniferous rocks of the Midland Valley of Scotland. *British Geological Survey Technical Report*, WA/96/29/R.

BUTTERWORTH, M A, and WILLIAMS, R W. 1958. The small spore floras of coals in the Limestone Coal Group and Upper Limestone Group of the Lower Carboniferous of Scotland. *Transaction of the Royal Society of Edinburgh*, Vol. 63, 353–392.

CAMERON, I B, FORSYTH, I H, HALL, I H S, and PEACOCK, J D. 1977. Sand and gravel resources of the Strathclyde Region of Scotland. *Report of the Institute of Geological Sciences*, No. 77/8.

CAMERON, I B, and STEPHENSON, D. 1985. *British regional geology: the Midland Valley of Scotland* (3rd edition). (London: HMSO for British Geological Survey.)

CAMPBELL, D. 1988. Detecting the dangers. *Surveyor*, October 1988, 14–15.

CHEEL, R J, and RUST, B R. 1982. Coarse-grained facies of glacio-marine deposits near Ottawa, Canada. 279–294 in *Research in Glacial, Glacio-fluvial and Glacio-lacustrine systems.* DAVIDSON-ARNOTT, R, NICKLING, W, and FAHEY, B D (editors). (Norwich: Geo Books.)

CHISHOLM, J I, and DEAN, J M. 1974. The Upper Old Red Sandstone of Fife and Kinross: a fluviatile sequence with evidence of a marine incursion. *Scottish Journal of Geology*, Vol. 10, 1–30.

CHISHOLM, J I, MCADAM, A D, and BRAND, P J. 1989. Lithostratigraphical classification of Upper Devonian and Lower Carboniferous rocks in the Lothians. *British Geological Survey Technical Report*, WA/89/26.

CLOUGH, C T, and seven others. 1911. The geology of the Glasgow district. *Memoir of the Geological Survey, Scotland.*

CONWAY, A, DENTITH, M C, DOODY, J J, and HALL, J. 1987. Preliminary interpretation of upper crustal structure across the Midland Valley of Scotland from two east–west seismic refraction profiles. *Journal of the Geological Society, London*, Vol. 144, 867–870.

COOMBS, D S, and WILKINSON, J F G. 1969. Lineages and fractionation trends in undersaturated volcanic rocks from the East Otago volcanic province (New Zealand) and related rocks. *Journal of Petrology*, Vol. 10, 440–501.

COTTON, W R. 1968. A geophysical survey of the Campsie and Kilpatrick hills. Unpublished PhD thesis, University of Glasgow.

CRAIG, P M. 1980. The volcanic geology of the Campsie Fells area, Stirlingshire. Unpublished PhD thesis, University of Lancaster.

CRAIG, P M, and HALL, I H S. 1975. The Lower Carboniferous rocks of the Campsie–Kilpatrick area. *Scottish Journal of Geology*, Vol. 11, 171–174.

CRAIG, R. 1873. On the glacial deposits of North Ayrshire and Renfrewshire. *Transactions of the Geological Society of Glasgow*, Vol. 4, 138–164.

CRAIG, R. 1877. Notes on a shell bed at the west end of Arkleston tunnel, near Paisley. *Transactions of the Geological Society of Glasgow*, Vol. 5, 332.

CURRIE, A. 1866. On the occurence of coal beds under the traps of the Bowling Hills. *Transactions of the Geological Society of Glasgow*, Vol. 2, 149–151.

CURRIE, E D. 1954. Scottish Carboniferous goniatites. *Transactions of the Royal Society of Edinburgh*, Vol. 62, 527–602.

DAVIDSON, K A S, SOLA, M, POWELL, D W, and HALL, J. 1984. Geophysical model for the Midland Valley of Scotland. *Transactions of the Royal Society of Edinburgh: Earth Sciences*, Vol. 75, 175–181.

DE SOUZA, H A F. 1979. The geochronology of Scottish Carboniferous volcanism. Unpublished PhD thesis, University of Edinburgh.

DEAN, M T. 1987. Carboniferous conodonts from the Lower and Upper Limestone Groups of the Scottish Midland Valley. Unpublished MPhil thesis, University of Nottingham.

DENTITH, M C, and HALL, J. 1989. MAVIS — an upper crustal seismic refraction experiment in the Midland Valley of Scotland. *Geophysical Journal International*, Vol. 99, 627–643.

DENTITH, M C, and HALL, J. 1990. MAVIS: geophysical constraints on the structure of the Carboniferous basin of West Lothian, Scotland. *Transactions of the Royal Society of Edinburgh: Earth Sciences*, Vol. 81, 117–126.

DENTITH, M C, TRENCH, A, and BLUCK, B J. 1992. Geophysical constraints on the nature of the Highland Boundary Fault Zone in western Scotland. *Geological Magazine*, Vol. 129, 411–419.

DEWEY, J F. 1982. Plate tectonics and the evolution of the British Isles. *Journal of Geological Sciences*, Vol. 139, 371–412.

DICKSON, J H, JARDINE, W G, and PRICE, R J. 1976. Three late-Devensian sites in west-central Scotland. *Nature, London*, Vol. 262, 43–44.

DICKSON, J H, and six others. 1978. Palynology, palaeomagnetism and radiometric dating of Flandrian marine and freshwater sediments of Loch Lomond. *Nature, London*, Vol. 274, 548–553.

DOODY, J J, SMYTHE, D K, and WATTS, D R. 1993. Geophysical investigation of the Blane valley Pleistocene deposits. *Scottish Journal of Geology*, Vol. 29, 201–204.

ELDER, S, MCCALL, R J S, NEAVES, W D, and PRINGLE, A K. 1935. The drumlins of Glasgow. *Transactions of the Geological Society of Glasgow*, Vol. 19, 285–287.

ELLIOT, C J. 1984. Palynology and palaeoenvironments of the BGS Mains of Kilmaronock Borehole, Loch Lomond, Scotland. Unpublished MSc thesis, City of London Polytechnic and the Polytechnic of North London.

ELLIOT, R W. 1985. Central Scotland mineral portfolio: resources of clay and mudstone for brickmaking. *Open-file report of the British Geological Survey.*

EVANS, C J, KIMBELL, G S, and ROLLIN, K E. 1988. *Hot dry rock potential in urban areas. Investigation of the Geothermal Potential of the UK.* (Keyworth, Nottingham: British Geological Survey.)

FORSYTH, I H. 1961. The succession between Plean No. 1 Limestone and No. 2 Marine Band in the Carboniferous of the east Glasgow area. *Transactions of the Geological Society of Glasgow*, Vol. 24, 213–234.

FORSYTH, I H. 1978a. The Lower Carboniferous sequence in the Howwood Syncline, Renfrewshire. *Bulletin of the Geological Survey of Great Britain*, No. 60, 1–8.

FORSYTH, I H. 1978b. The Glenboig Marine Band in the Upper Limestone Group of the Namurian of central Scotland. *Bulletin of the Geological Survey of Great Britain*, No. 60, 17–22.

FORSYTH, I H. 1978c. The lower part of the Limestone Coal Group in the Glasgow district. *Report of the Institute of Geological Sciences*, No. 78/29.

FORSYTH, I H. 1979. The Lower Coal Measures of central Glasgow. *Report of the Institute of Geological Sciences*, No. 79/4.

FORSYTH, I H. 1980. The *Lingula* bands in the upper part of the Limestone Coal Group (E$_1$ Stage of the Namurian) in the Glasgow district. *Report of the Institute of Geological Sciences*, No. 79/16.

FORSYTH, I H. 1982. The stratigraphy of the Upper Limestone Group (E$_1$ and E$_2$ stages of the Namurian) in the Glasgow district. *Report of the Institute of Geological Sciences*, No. 82/4.

FORSYTH, I H. 1993. The stratigraphy of the central coalfield of Scotland. *British Geological Survey Technical Report*, WA/93/51R.

FORSYTH, I H, and BRAND, P J. 1986. Stratigraphy and stratigraphical palaeontology of Westphalian B and C in the Central Coalfield of Scotland. *Report of the British Geological Survey*, Vol. 18, No. 4.

FORSYTH, I H, HALL, I H S, and MCMILLAN, A A. 1996. Geology of the Airdrie district. *Memoir of the British Geological Survey*, Sheet 31W (Scotland).

FORSYTH, I H, and READ, W A. 1962. The correlation of the Limestone Coal Group above the Kilsyth Coking Coal in the Glasgow–Stirling region. *Bulletin of the Geological Survey of Great Britain*, No. 19, 29–52.

FORSYTH, I H, and WILSON, R B. 1965. Recent sections in the Lower Carboniferous of the Glasgow area. *Bulletin of the Geological Survey of Great Britain*, No. 22, 65–79.

FRANCIS, E H, FORSYTH, I H, READ, W A, and ARMSTRONG, M. 1970. The geology of the Stirling district. *Memoir of the Geological Survey of Great Britain*, Sheet 39 (Scotland).

FRASER, W. 1873. On some recently exposed sections in the Paisley clay-beds, and their relation to the Glacial Period. *Transactions of the Geological Society of Glasgow*, Vol. 4, 178–181.

FRYBERGER, S G, and SCHENK, C J. 1988. Pin-stripe lamination: a distinctive feature of modern and ancient aeolian sediments. *Sedimentary Geology*, Vol. 55, 1–15.

GEIKIE, A. 1863. On the phenomena of the glacial drift of Scotland. *Transactions of the Geological Society of Glasgow*, Vol. 1, Pt 2, 1–190.

GEIKIE, J. 1872. *On changes of climate during the Glacial Epoch.* (London: Trubner & Co.)

GEORGE, T N. 1974. Prologue to a geomorphology of Britain. *Institute of British Geographers Special Publication*, No. 7, 113–125.

GREGORY, J W. 1914. Corries, with special reference to those of the Campsie Fells. *Transactions of the Geological Society of Glasgow*, Vol. 15, 84–98.

HALL, D H, and DAGLEY, P. 1970. Regional magnetic anomalies: an analyses of the smoothed Aeromagnetic Map of Great Britain and Northern Ireland. *Report of the Institute of Geological Sciences*, No. 70/10.

HALL, I H S, and CHISHOLM, J I. 1987. Aeolian sediments in the late Devonian of the Scottish Midland Valley. *Scottish Journal of Geology*, Vol. 23, 203–208.

HALL, J. 1971. A preliminary seismic survey adjacent to the Rashiehill borehole near Slamannan, Stirlingshire. *Scottish Journal of Geology*, Vol. 7, 170–174.

HALL, J. 1974. A seismic reflection survey of the Clyde Plateau Lavas in North Ayrshire and Renfrewshire. *Scottish Journal of Geology*, Vol. 9, 253–279.

HAMILTON, J. 1956. Mineralogy of basalts from the western Kilpatrick Hills and its bearing on the petrogenesis of the Scottish Carboniferous olivine basalts. *Transactions of the Edinburgh Geological Society*, Vol. 16, 280–298.

HASZELDINE, R S. 1988. Crustal lineaments in the British Isles: their relationship to Carboniferous basins. 53–68 in *Sedimentation in a synorogenic basin complex: the Upper Carboniferous of northwest Europe.* BESLY, B M, and KELLING, G (editors). (Glasgow: Blackie and Son.)

HINXMAN, L W, ANDERSON, E M, and CARRUTHERS, R G. 1920. The economic geology of the Central Coalfield of Scotland, Area IV, Paisley, Barrhead and Renfrew. *Memoir of the Geological Survey, Scotland.*

HUTTON, A N. 1965. Foraminifera of the Upper Limestone Group of the Scottish Carboniferous. Unpublished PhD thesis, University of Glasgow.

JACK, R L. 1875. Notes on a till or boulder clay with broken shells, in the lower valley of the River Endrick, near Loch Lomond, and its relation to certain other glacial deposits. *Transactions of the Geological Society of Glasgow*, Vol. 5, 5–25.

JAMIESON, T F. 1865. On the history of the last geological changes in Scotland. *Journal of the Geological Society of London*, Vol. 21, 161–203.

JARDINE, W G. 1965. Note on a temporary exposure in central Glasgow of Quaternary sediments with slump and load structures. *Scottish Journal of Geology*, Vol. 1, 221–224.

JARDINE, W G. 1969. Quaternary deposits near Garscadden Mains, Glasgow. *Proceedings of the Geological Society of Glasgow*, Session 110, 51–53.

JARDINE, W G. 1977. The Quaternary marine record in southwest Scotland and the Scottish Hebrides. 99–118 in *The Quaternary history of the Irish Sea.* KIDSON, C, and TOOLEY, M J (editors). (Liverpool: Seel House Press.)

JARDINE, W G (editor). 1980. *Glasgow region: field guide.* (Glasgow: Quaternary Research Association.)

JARDINE, W G. 1986. The geological and geomorphological setting of the estuary and Firth of Clyde. *Proceedings of the Royal Society of Edinburgh*, Vol. 90B, 25–41.

JARDINE, W G, and five others. 1988. A late Middle Devensian interstadial site at Sourlie, near Irvine, Strathclyde. *Scottish Journal of Geology*, Vol. 24, 288–295.

JARDINE, W G, and MOISLEY, H A. 1967. Note on a temporary exposure of Quaternary deposits at Scotstoun House, Glasgow. *Proceedings of the Geological Society of Glasgow*, Session 108, 25–27.

KENNEDY, W Q. 1958. Tectonic evolution of the Midland Valley of Scotland. *Transactions of the Geological Society of Glasgow*, Vol. 23, 107–133.

LASKEY, CAPTAIN. 1822. Notice in regard to marine shells found in the line of the Ardrossan Canal. *Memoirs of the Wernerian Natural History Society*, Vol. 4, 568–569.

LEEDER, M R. 1982. Upper Palaeozoic basins of the British Isles–Caledonian inheritance versus Hercynian plate marginal processes. *Journal of the Geological Society, London*, Vol. 139, 479–481.

LEEDER, M R, and MCMAHON, A H. 1988. Upper Carboniferous (Silesian) basin subsidence in northern Britain. 43–52 in *Sedimentation in a synorogenic basin complex: the Upper*

Carboniferous of northwest Europe. BESLY, B M, and KELLING, G (editors). (Glasgow: Blackie and Son.)

LINTON, D L. 1951. Problems of Scottish scenery. *Scottish Geographical Magazine*, Vol. 67, 65–85.

MACDONALD, J G, and WHYTE, F. 1981. Petrochemical evidence for the genesis of a Lower Carboniferous transitional basaltic suite in the Midland Valley of Scotland. *Transactions of the Royal Society of Edinburgh: Earth Sciences*, Vol. 72, 75–88.

MACDONALD, R. 1975. Petrochemistry of the early Carboniferous (Dinantian) lavas of Scotland. *Scottish Journal of Geology*, Vol. 11, 269–314.

MACDONALD, R, THOMAS, J E, and RIZZELLO, S A. 1977. Variations in basalt chemistry with time in the Midland Valley province during the Carboniferous and Permian. *Scottish Journal of Geology*, Vol. 13, 11–22.

MACFARLANE, J. 1858. Memorandum of shells and a deer's horn found in a cutting of the Forth and Clyde Junction Railway, Dunbartonshire. *Proceedings of the Royal Physical Society*, Vol. 1, 163–165.

MACGREGOR, A G. 1928. The classification of Scottish Carboniferous olivine-basalts and mugearites. *Transactions of the Geological Society of Glasgow*, Vol. 18, 324–360.

MACGREGOR, A G. 1960. Divisions of the Carboniferous on Geological Survey Scottish maps. *Bulletin of the Geological Survey of Great Britain*, No. 16, 127–130.

MACGREGOR, M, DINHAM, C H, BAILEY, E B, and ANDERSON, E M. 1925. The geology of the Glasgow district (2nd edition). *Memoir of the Geological Survey of Great Britain*.

MACHETTE, M N. 1985. Calcic soils in the southwestern United States. 1–21 *in* Soils and Quaternary Geology of the southwestern United States. WEIDE, D L (editor). *Geological Society of America Special Paper*, No. 203.

MACKIEWICZ, N E, POWELL, R D, CARLSON, P R, and MOLNIA, B F. 1984. Interlaminated ice-proximal glacimarine sediments in Muir Inlet, Alaska. *Marine Geology*, Vol. 57, 113–147.

MAXWELL, G M. 1971. The geophysical investigation of sub-surface hazards due to abandoned coal-mines. Unpublished PhD thesis, University of Strathclyde.

MCGOWN, A, and MILLER, D. 1984. Stratigraphy and properties of the Clyde alluvium. *Quarterly Journal of Engineering Geology*, Vol. 17, 243–258.

MCLINTOCK, W F P, and PHEMISTER, J. 1929. A gravitational survey over the buried Kelvin valley at Drumry, near Glasgow. *Transactions of the Royal Society of Edinburgh*, Vol. 56, 141–156.

MCMILLAN, A A, and BROWNE, M A E. 1989. Fold basins in Late-Devensian glacimarine sediments at Shieldhall, Glasgow. *Scottish Journal of Geology*, Vol. 25, 295–305.

MENZIES, J. 1976. The glacial geomorphology of Glasgow with particular reference to the drumlins. Unpublished PhD thesis, University of Edinburgh.

MENZIES, J. 1981. Investigations into the Quaternary deposits and bedrock topography of central Glasgow. *Scottish Journal of Geology*, Vol. 17, 155–168.

MERRITT, J W, and ELLIOT, R W. 1984. Central Scotland mineral portfolio: hard rock aggregate resources. *Open-file Report of the British Geological Survey*.

MITCHELL, G F. 1952. Late-Glacial deposits at Garscadden Mains, near Glasgow. *New Phytology*, Vol. 50, 277–286.

MOORE, E W J. 1939. The goniatite genus *Dimorphoceras* and its development in the British Carboniferous. *Proceedings of the Yorkshire Geological Society*, Vol. 24, 103–128.

MORTON, D J. 1979. Palaeogeographical evolution of the Lower Old Red Sandstone basin in the western Midland Valley. *Scottish Journal of Geology*, Vol. 15, 97–116.

NEILSON, J. 1896. On the occurrence of marine organisms in the boulder clay of the Glasgow district. *Transactions of the Geological Society of Glasgow*, Vol. 10, 273–279.

NEVES, R, GUEINN, K J, CLAYTON, G, IOANNIDES, N S, and NEVILLE, R S W. 1972. A scheme of miospore zones for the British Dinantian. *Compte Rendu 7e Congrès International de Stratigraphie et de Géologie Carbonifère, Krefeld 1971*, Vol. 1, 347–353.

NEVES, R, READ, W A, and WILSON, R B. 1965. Note on recent spore and goniatite evidence from the Passage Group, of the Scottish Upper Carboniferous succession. *Scottish Journal of Geology*, Vol. 1, 185–188.

NICOLL, J C. 1990. Summerston landfill site: containment works — a case study. *Preprints of the Institution of Water and Environment Management Conference 1990*, 16.1–16.15.

NORTH AMERICAN COMMISSION ON STRATIGRAPHIC NOMENCLATURE. 1983. North American Stratigraphic Code. *American Association of Petrology Geological Bulletin*, Vol. 67, 841–875.

PALMER, J A, PERRY, S P G, and TARLING, D H. 1985. Carboniferous magnetostratigraphy. *Journal of the Geological Society, London*, Vol. 142, 945–955.

PAPROTH, E, FEIST, R, and FLAJS, G. 1991. Decision on the Devonian–Carboniferous boundary stratotype. *Episodes*, Vol. 14, 331–336.

PATERSON, I B, and HALL, I H S. 1986. Lithostratigraphy of the late Devonian and early Carboniferous rocks in the Midland Valley of Scotland. *Report of the British Geological Survey*, No. 18/3.

PATERSON, I B, HALL, I H S, and STEPHENSON, D. 1990. Geology of the Greenock district. *Memoir of the British Geological Survey*, Sheets 30W and part of 29E (Scotland).

PATERSON, I B, MCADAM, A D, and MACPHERSON, K A T. In press. Geology of the Hamilton district. *Memoir of the British Geological Survey*, Sheet 23W (Scotland).

PEACH, A M. 1909. Boulder distribution from Lennoxtown, Scotland. *Geological Magazine*, Vol. 46, 26–31.

PEACOCK, J D. 1971. Marine shell radiocarbon dates and the chronology of deglaciation in western Scotland. *Nature, London*, Vol. 230, 43–45.

PEACOCK, J D. 1975. Scottish late and post-glacial marine deposits. 45–48 *in Quaternary studies in north east Scotland*. GEMMELL, A M D (editor). (Aberdeen: University of Aberdeen.)

PEACOCK, J D. 1981. Scottish Late-Glacial marine deposits and their environmental significance. 222–236 *in The Quaternary in Britain*. NEALE, J, and FLENLEY, J (editors). (Oxford and New York: Pergamon Press.)

PEACOCK, J D. 1989. Marine molluscs and late Quaternary environmental studies with particular reference to the Late-Glacial Period in northwest Europe: a review. *Quaternary Science Reviews*, Vol. 8, 179-192.

PEACOCK, J D, GRAHAM, D K, and WILKINSON, I P. 1978. Late-Glacial and post-Glacial marine environments at Ardyne, Scotland and their significance in the interpretation of the history of the Clyde sea area. *Report of the Institute of Geological Sciences*, No. 78/17.

PEDLEY, R C. 1991. *Interactive 2.5D gravity and magnetic modelling program. User manual, Integrated Geophysical Services.* (Keyworth, Nottingham: British Geological Survey.)

PENN, I E, SMITH, I F, and HOLLOWAY, S. 1984. *Interpretation of a deep seismic reflection profile in the Glasgow area. Investigation of the geothermal potential of the UK.* (Keyworth, Nottingham: British Geological Survey.)

PIPER, D J W, LETSON, J R J, DELORE, A M, and BARRIE, C Q. 1983. Sediment accumulation in low-sedimentation, wave-dominated, glaciated inlets. *Sedimentary Geology*, Vol. 36, 195–215.

POWELL, D W. 1963. Significance of differences in magnetization along certain dolerite dykes. *Nature, London*, Vol. 199, 674–676.

POWELL, D W. 1970. Magnetised rocks within the Lewisian of Western Scotland and under the Southern Uplands. *Scottish Journal of Geology*, Vol. 6, 353–369.

POWELL, D W. 1978. Gravity and magnetic anomalies attributable to basement sources under northern Britain. 107–114 in Crustal evolution in northwestern Britain and adjacent regions. BOWES, D R, and LEAKE, B E (editors). *Special Issue of the Geological Journal*, No. 10.

PRICE, R J. 1983. *Scotland's environment during the last 30 000 years.* (Edinburgh: Scottish Academic Press.)

QURESHI, I R. 1970. A gravity survey of a region of the Highland Boundary Fault in Scotland. *Quarterly Journal of the Geological Society of London*, Vol. 125, 481–502.

RAMSBOTTOM, W H C. 1977. Correlation of the Scottish Upper Limestone Group (Namurian) with that of the north of England. *Scottish Journal of Geology*, Vol. 13, 327–330.

RAMSBOTTOM, W H C, and six others. 1978. A correlation of Silesian rocks in the British Isles. *Geological Society of London Special Report*, No. 10.

READ, W A. 1988. Controls on Silesian sedimentation in the Midland Valley of Scotland. 222–241 in *Sedimentation in a synorogenic basin complex: the Upper Carboniferous of northwest Europe.* BESLY, B M, and KELLING, G (editors). (Glasgow and London: Blackie and Son.)

READ, W A. 1989. The interplay of sedimentation, volcanicity and tectonics in the Passage Group (Arnsbergian, E₂ to Westphalian A) in the Midland Valley of Scotland. 143–152 in The role of tectonics in Devonian and Carboniferous sedimentation in the British Isles. ARTHURTON, R J, GUTTERIDGE, P, and NOLAN, S C (editors). *Occasional Publication of the Yorkshire Geological Society*, No. 6.

READ, W A, and JOHNSON, S R H. 1967. The sedimentology of sandstone formations within the Upper Old Red Sandstone and lowest Calciferous Sandstone Measures, west of Stirling, Scotland. *Scottish Journal of Geology*, Vol. 3, 242–267.

RICHEY, J E, ANDERSON, E M, and MACGREGOR, A G. 1930. The geology of north Ayrshire. *Memoir of the Geological Survey, Scotland*.

ROBERTSON, D, and CROSSKEY, H W. 1874. On the Post-Tertiary fossiliferous beds of Scotland. XVI — Stobcross. *Transactions of the Geological Society of Glasgow*, Vol. 4, 245–251.

ROBERTSON, T, and HALDANE, D. 1937. The economic geology of the Central Coalfield, Area I, Kilsyth and Kirkintilloch. *Memoir of the Geological Survey, Scotland*.

ROBINS, N S. 1990. Hydrogeology of Scotland. (London: HMSO for the British Geological Survey.)

ROLFE, W D I. 1966. Woolly rhinoceros from the Scottish Pleistocene. *Scottish Journal of Geology*, Vol. 2, 253–258.

ROLLIN, K E. 1987a. *Catalogue of geothermal data for the land area of the United Kingdom. Third revision: April 1987. Investigation of the geothermal potential of the UK.* (Keyworth, Nottingham: British Geological Survey.)

ROLLIN, K E. 1987b. The geothermal environment in Scotland. *Modern Geology*, Vol. 11, 235–250.

ROSE, J. 1975. Raised beach gravels and ice-wedge casts at Old Kilpatrick, near Glasgow. *Scottish Journal of Geology*, Vol. 11, 15–21.

ROSE, J. 1981. Field guide to the Quaternary geology of the southeastern part of the Loch Lomond basin. *Proceedings of the Geological Society of Glasgow*, Sessions 122/123, 12–28.

ROSE, J. 1989. Stadial type sections in the British Quaternary. 45–67 in *Quaternary type sections: imagination or reality?* ROSE, J. and SCHLUCHTER, C (editors). (Rotterdam: Balkema.)

ROSE, J. LOWE, J J, and SWITSUR, R. 1988. A radiocarbon date on plant detritus beneath till from the type area of the Loch Lomond Readvance. *Scottish Journal of Geology*, Vol. 24, 113–124.

RUSSELL, M J. 1971. North–south geofractures in Scotland and Ireland. *Scottish Journal of Geology*, Vol. 8, 75–84.

RUST, B R, and ROMANELLI, R. 1975. Late Quaternary sub-aqueous outwash deposits near Ottawa, Canada. 177–192 in Glaciofluvial and glaciolacustrine sedimentation. JOPLING, A V, and MCDONALD, B S (editors). *Society of Economic Palaeontologists and Mineralogists Special Publication*, No. 23.

SABINE, P A, GUPPY, E M, and SERGEANT, G A. 1969. Geochemistry of sedimentary rocks. 1. Petrograph and chemistry of arenaceous rocks. *Report of the Institute of Geological Sciences*, No. 69/1.

SAUNDERSON, H C. 1977. The sliding bed facies in esker sands and gravels: a criterion for full-pipe (tunnel) flow? *Sedimentology*, Vol. 24, 623–638.

SCOTT, A C, GALTIER, J, and CLAYTON, G. 1984. Distribution of anatomically preserved floras in the Lower Carboniferous in Western Europe. *Transactions of the Royal Society of Edinburgh: Earth Sciences*, Vol. 75, 311–340.

SCOTT, W B. 1986. Nodular carbonates in the Lower Carboniferous, Cementstone Group of the Tweed Embayment, Berwickshire: evidence for a former sulphate evaporite facies. *Scottish Journal of Geology*, Vol. 22, 325–345.

SHAKESBY, R A. 1976. The Lennoxtown essexite erratics train, central Scotland. Unpublished PhD thesis, University of Edinburgh.

SHEARMAN, D J, MOSSOP, G, DUNSMORE, H, and MARTIN, M. 1972. Origin of gypsum veins by hydraulic fracture. *Transactions of the Institute of Mining and Metallurgy, Section B*, Vol. 81, 149–155.

SIMPSON, J B. 1933. The late-glacial readvance moraines of the Highland border west of the River Tay. *Transactions of the Royal Society of Edinburgh*, Vol. 57, 633–645.

SISSONS, J B. 1966. Relative sea-level changes between 10 300 and 8300 BP in part of the Carse of Stirling. *Transactions of the Institute of British Geographers*, Vol. 39, 19–29.

SISSONS, J B. 1969. Drift stratigraphy and buried morphological features in the Grangemouth–Falkirk–Airth area, central Scotland. *Transactions of the Institute of British Geographers*, Vol. 48, 19–50.

SISSONS, J B. 1976. *Scotland.* The geomorphology of the British Isles. (London: Methuen.)

SISSONS, J B, and BROOKS, C L. 1971. Dating of early postglacial land and sea level changes in the western Forth valley. *Nature Physical Science*, Vol. 234, 124–127.

SMEDLEY, P L. 1986. The relationship between calc-alkaline volcanism and within-plate continental rift volcanism: evidence from Scottish Palaeozoic lavas. *Earth and Planetary Science Letters*, Vol. 77, 113–128.

SMEDLEY, P L. 1988. Trace element and isotope variations in Scottish and Irish Dinantian volcanism: evidence for an OIB-like mantle source. *Journal of Petrology*, Vol. 29, 413–443.

SMITH, A H V, and BUTTERWORTH, M A. 1967. Miospores in the coal seams of the Carboniferous of Great Britain. *Special Papers in Palaeontology*, No. 1.

SMITH, J. 1836. On changes of the levels of sea and land in the west of Scotland. *Proceedings of the Geological Society of London*, Vol. 2, 427–429.

SMITH, J. 1838. On the last changes in the relative levels of the land and sea in the British Isles. *Memoirs of the Wernerian Natural History Society*, Vol. 8, 49–113.

SMITH, J. 1862. *Researches in newer Pliocene and post-Tertiary geology*. (Glasgow: John Gray.)

STEDMAN, C. 1988. Namurian E_1 tectonics and sedimentation in the Midland Valley of Scotland: rifting versus strike-slip influence. 242–254 in *Sedimentation in a synorogenic basin complex: the Upper Carboniferous of northwest Europe*. BESLY, B M, and KELLING, G (editors). (Glasgow and London: Blackie and Son.)

STORETVEDT, K M, DEUTSCH, E R, and MURTHY, G S. 1991. On the British Siluro-Devonian palaeomagnetic field problem: reply to comments by T H Torsvik et al. *Geophysical Journal International*, Vol. 105, 475–476.

SUTHERLAND, D G. 1984. The Quaternary deposits and landforms of Scotland and the neighbouring shelves: a review. *Quaternary Science Reviews*, Vol. 3, 157–254.

TAIT, A M. 1973. Sedimentation of the Craigmaddie Sandstone Formation, western Midland Valley of Scotland. Unpublished PhD thesis, University of Glasgow.

THOMSON, T. 1835. On a deposit of recent marine shells at Dalmuir, Dunbartonshire. *Records of General Science*, Vol. 1, 131–135.

TORSVIK, T H. 1985. Magnetic properties of the Lower Old Red Sandstone lavas in the Midland Valley, Scotland; palaeomagnetic and tectonic considerations. *Physics of the Earth and Planetary Interiors*, Vol. 39, 194–207.

TORSVIK, T H, LYSE, O, ATTERAS, G, and BLUCK, B J. 1989. Palaeozoic palaeomagnetic results from Scotland and their bearing on the British apparent polar wander path. *Physics of the Earth and Planetary Interiors*, Vol. 55, 93–105.

UPTON, B J G, ASPEN, P, and CHAPMAN, N A. 1983. The upper mantle and deep crust beneath the British Isles: evidence from inclusions in volcanic rocks. *Journal of the Geological Society, London*, Vol. 140, 105–121.

WALDERHAUG, H J, TORSVIK, T H, and LOVLIE, R. 1991. Experimental CRM production in a basaltic rock; evidence for stable, intermediate palaeomagnetic directions. *Geophysical Journal International*, Vol. 105, 747-756.

WALLACE, J. 1902. Notes on a recent excavation in Sauchiehall Street. *Transactions of the Geological Society of Glasgow*, Vol. 12, 79–80.

WALLACE, J. 1905. Notes on excavations in Glasgow. *Transactions of the Geological Society of Glasgow*, Vol. 12, 220–222.

WATTANANIKORN, K. 1978. The determination of the nature of the reserves of typical igneous intrusive rock quarries by both field and laboratory geophysical measurements. Unpublished PhD thesis, University of Strathclyde.

WHYTE, F. 1968. Lower Carboniferous volcanic vents in the west of Scotland. *Bulletin Volcanique*, Vol. 32, 253–268.

WHYTE, F, and MACDONALD, J G. 1974. Lower Carboniferous vulcanicity in the northern part of the Clyde plateau. *Scottish Journal of Geology*, Vol. 10, 187–198.

WILSON, R B. 1966. A study of the Neilson Shell Bed, a Scottish Lower Carboniferous marine shale. *Bulletin of the Geological Survey of Great Britain*, No. 24, 105–130.

WILSON, R B. 1967. A study of some Namurian marine faunas of central Scotland. *Transactions of the Royal Society of Edinburgh*, Vol. 66, 445–490.

WILSON, R B. 1989. A study of some Dinantian marine macrofossils of central Scotland. *Transactions of the Royal Society of Edinburgh: Earth Sciences*, Vol. 80, 91–126.

WRIGHT, J. 1896. Boulder-clay, a marine deposit. *Transactions of the Geological Society of Glasgow*, Vol. 10, 263–272.

YOUNG, J. 1873a. On the occurrence of the remains of Carboniferous fishes in certain thin beds of indurated shale lying between sheets of trap, composing the Kilpatrick range of hills at Auchentorlie, near Bowling. *Transactions of the Geological Society of Glasgow*, Vol. 4, 77.

YOUNG, J. 1873b. Notes on a section of strata containing beds of impure coal and plant remains showing structure at Glenarbuck near Bowling. *Transactions of the Geological Society of Glasgow*, Vol. 4, 123–128.

APPENDIX 1

List of BGS boreholes in the Glasgow district and adjacent areas that are cited in the text

Name	Accession number in BGS files	Grid reference
Solid		
Barnhill	NS 47NW/2	4269 7571
Glenburn	NS 46SE/164	4783 6066
Hurlet	NS 56SW/333	5111 6123
Kipperoch	NS 37NE/20	3727 7742
Lawmuir	NS 57SW/161,2	4592 7555
Maryhill*	NS 56NE/1735	5718 6856

*Published, internal report NL 83/2.
Outline stratigraphy of remaining boreholes has been published in IGS Reports Nos. 78/21; 79/12; 81/11.

Name	Accession number	Grid reference
Drift		
Bridgeton	NS 66SW/826	6121 6367
Erskine Bridge	NS 47SE/18	4634 7251
Gartness	NS 48NE/2	4974 8671
Killearn	NS 58SW/3	5100 8467
Law	NS 85SW/436	8357 5247
Linwood	NS 46NW/62	4459 6588
Mains of Kilmaronock	NS 48NW/3	4483 8829

Published, British Geological Survey Research Report SA/89/1

APPENDIX 2

1:10 000 maps (Solid)

The maps at 1:10 000 scale covering, wholly or in part, the solid rocks in 1:50 000 Sheet 30E are listed below with the names of the surveyors (P M Craig, I H Forsyth, I H S Hall, J M Dean and I B Paterson) and the date of survey.

The maps are not published but are available for consultation in the Library of the British Geological Survey, Murchison House, West Mains Road, Edinburgh EH9 3LA. Dyeline copies can be purchased from the Bookshop.

National Grid sheet

NS 45 NW	Stephenson	1979
NS 45 NE	Stephenson	1979
NS 46 NW	Paterson and Forsyth	1980–84
NS 46 NE	Paterson and Forsyth	1980–84
NS 46 SW	Paterson, Stephenson and Forsyth	1979–84
NS 46 SE	Paterson, Stephenson and Forsyth	1979–84
NS 47 NW	Hall	1976–78
NS 47 NE	Hall	1974–76
NS 47 SW	Hall and Paterson	1977–80
NS 47 SE	Hall and Paterson	1977–79
NS 48 NW	Hall	1979–80
NS 48 NE	Hall	1979
NS 48 SW	Hall	1977–79
NS 48 SE	Hall	1974–77
NS 55 NW	Forsyth	1977
NS 55 NE	Forsyth and Craig	1968–77
NS 56 NW	Forsyth	1984
NS 56 NE	Forsyth and Hall	1973–84
NS 56 SW	Forsyth and Dean	1974–87
NS 56 SE	Forsyth and Dean	1973–87
NS 57 NW	Hall	1974
NS 57 NE	Hall, Craig and Forsyth	1953–78
NS 57 SW	Hall	1977–78
NS 57 SE	Forsyth and Hall	1953–73
NS 58 NW	Hall	1979
NS 58 NE	Craig	1974–75
NS 58 SW	Hall and Craig	1974–75
NS 58 SE	Craig	1974–75
NS 65 NW	Forsyth and Craig	1968–80
NS 66 NW	Forsyth	1953–54
NS 66 SW	Forsyth	1954–76
NS 67 NW	Forsyth and Craig	1953–72
NS 67 SW	Forsyth	1953–75
NS 68 NW	Craig	1971–72
NS 68 SW	Craig	1970

APPENDIX 3

1:10 000 maps (Drift)

The maps at 1:10 000 scale covering, wholly or in part, the superficial deposits in 1:50 000 Sheet 30E are listed below with the names of the surveyors (A M Aitken, M A E Browne, P M Craig, J M Dean, I H Forsyth, I H S Hall, A A McMillan, S K Monro, I B Paterson and D Stephenson) and the date of survey.

The maps are not published but are available for consultation in the Library, the British Geological Survey, Murchison House, West Mains Road, Edinburgh EH9 3LA. Dyeline copies can be purchased from the Bookshop.

National Grid sheet

NS 45 NW	Stephenson, Monro	1978-79
NS 45 NE	Stephenson	1979
NS 46 NW	Browne, Paterson and Forsyth	1977–84
NS 46 NE	Browne, Paterson and Forsyth	1977–84
NS 46 SW	Browne, Paterson, Stephenson and Forsyth	1977–84
NS 46 SE	Browne, Paterson, Stephenson and Forsyth	1977–84
NS 47 NW	Hall	1976–78
NS 47 NE	Hall	1974–76
NS 47 SW	Browne, Hall and Paterson	1977–80
NS 47 SE	Browne, Hall, Paterson and McMillan	1977–80
NS 48 NW	Hall, Browne and Aitken	1979–84
NS 48 NE	Hall and Aitken	1979–84
NS 48 SW	Hall and Aitken	1979–84
NS 48 SE	Hall and Aitken	1974–84
NS 55 NW	Browne and Forsyth	1977
NS 55 NE	Browne, Craig and Forsyth	1968–77
NS 56 NW	Browne, Hall and McMillan	1978–84
NS 56 NE	Browne, Forsyth, Hall and McMillan	1978–84
NS 56 SW	Browne and Dean	1974–84
NS 56 SE	Browne and Dean	1973–84
NS 57 NW	Hall	1974
NS 57 NE	Forsyth, Craig and Hall	1953–78
NS 57 SW	Browne, Hall and McMillan	1976–84
NS 57 SE	Forsyth, Hall and McMillan	1953–84
NS 58 NW	Hall and Aitken	1979–84
NS 58 NE	Craig and Aitken	1974–84
NS 58 SW	Craig, Hall and Aitken	1974–84
NS 58 SE	Craig and Aitken	1974–84
NS 65 NW	Browne, Craig and Forsyth	1968–80
NS 66 NW	Browne and McMillan	1975–83
NS 66 SW	Forsyth, Browne and McMillan	1954–83
NS 67 NW	Forsyth and Craig	1953–72
NS 67 SW	Forsyth and McMillan	1953–80
NS 68 NW	Craig	1971–72
NS 68 SW	Craig	1970–72

APPENDIX 4

List of Geological Survey photographs

Copies of these photographs are deposited for public reference in the Library of the British Geological Survey, West Mains Road, Edinburgh, EH9 3LA. The photographs belong to the Series C and D as indicated. Prints can be purchased from the Bookshop.

C 2101-2	Alternating beds of mudstone and cementstone of the Ballagan Formation, Ballagan Burn, Strathblane.
C 2103-4	South scarp of the Strathblane Hills showing trap features of the Clyde Plateau Volcanic Formation.
C 2105-6	Volcanic vents of Lower Carboniferous age filled with agglomerate and basalt, near Strathblane.
C 2107	U-shaped valley between the volcanic vents of Dumgoyne and Dumfoyne, near Strathblane.
C 2108	Landslip and scree on the flanks of Dumgoyne, volcanic vent of Lower Carboniferous age, near Strathblane.
C 2109	Dumfoyne, volcanic vent of Lower Carboniferous age, near Strathblane.
C2110-11	North scarp of Campsie Fells formed by a line of Lower Carboniferous vents.
C 2112	Columnar basalt in Lower Carboniferous vent on north scarp of Campsie Fells.
C 2113-15	Corrie of Balglass cut in north scarp of Campsie Fells with landslips (left) and terminal moraine.
C 2123-4	Hill peat on the Campsie Fells being eroded.
C 2417-8	Stoop and room mining in the Bishopbriggs Sandstone (Upper Limestone Formation) Huntershill, Bishopbriggs, north Glasgow.
C 2419	Section of Wilderness Till Formation, Hamilton Hill Quarry, Glasgow.
C 2420	Erosive surface in Cowlairs Sandstone (Limestone Coal Formation) in railway cutting west of Possilpark Station, Glasgow.
C 3093-5	Torsion balance used to survey buried channel of River Kelvin.
C 3120	Western scarp of Kilpatrick Hills formed by lavas of the Clyde Plateau Volcanic Formation.
C 3132-4	Crag and tail, Duncryne Hill. Crag is formed of Lower Carboniferous vent and tail is composed of Gartocharn Till Formation.
C 3135-8	Terminal moraine of Loch Lomond Stade age, Cameron Muir.
C 3554	Teschenite sill and underlying sedimentary rocks, High Craig Quarry, Johnstone.
C 3555-7	Stumps of fossil trees below sill, High Craig Quarry, Johnstone.
C 3560-1	Fossil Grove with casts of rooted *Lepidodendron* trees, Victoria Park, Glasgow.
C 3589-95	Shallow workings in Main Coal (Limestone Coal Formation) showing stoops and rooms, Thornliebank.
C 3953-5	Bedded tuff of the Clyde Plateau Volcanic Formation, Greenland Reservoir.
C 3956	Long Craigs, scarp at western end of Kilpatrick Hills formed by lavas of Clyde Plateau Volcanic Formation.
C 3957	Haw Crag, fault controlled lava scarp, in the Kilpatrick Hills, near Bowling.
C 3958	Faulting in lavas of the Clyde Plateau Volcanic Formation, picked out by gullies, near Bowling.
C 3959	Dip and scarp features in lavas of the Clyde Plateau Volcanic Formation, Greenside Reservoir.
C 3964	Fault controlled scarp slopes formed by lavas of the Clyde Plateau Volcanic Formation, Hill of Dun.
C3965	Drumlin, composed of Wilderness Till Formation, Golden Hill, Duntocher.
C 4041-5	Shallow workings in the Knightswood Gas Coal (Limestone Coal Formation) showing stoops, rooms and collapsed wastes, Glasgow University.
C 4075-6	Index Limestone and Bishopbriggs Sandstone (Upper Limestone Formation) in railway cutting, Bishopbriggs.
C 4081-2	Stoop and room workings in Baldernock Limestone (Lawmuir Formation), Linn of Baldernock.
C 4087	Unconformity at base of Bishopbriggs Sandstone (Upper Limestone Formation), South Coltpark Quarry, Bishopbriggs.
C 4088	Mudstone with ironstone beds (Lower Limestone Formation) used for bricks, Blairskaith Quarry, Balmore.
C 4240	Faulted mudstone against sandstone and fireclay (Upper Limestone Formation), Garngad Quarry.
D 98	Crag and Tail, Duncryne Hill. Crag is form by Lower Carboniferous vent and tail is composed of Gartocharn Till Formation.
D 99	General view of Campsie Fells showing north and south-west scarps formed by lavas and vent of the Clyde Plateau Volcanic Formation.
D 100	Lower Carboniferous vent, cutting Upper Devonian strata, Dumgoyach, Strathblane.
D 101	Gorge cut in Upper Devonian strata, Finnich Glen.
D 102	Quartz-conglomerate (Lawmuir Formation), Muirhouse, Strathblane.
D 103	Dumgoyne, volcanic vent of Lower Carboniferous age, near Strathblane.
D 104	Dumgoyne and Dumfoyne, volcanic vents of Lower Carboniferous age, near Strathblane.
D 105	Ballagan Glen, formed where erosion has cut through the overlying Clyde Plateau Volcanic Formation to expose Ballagan Formation.
D 106-8	Multi-phase columnar jointing in basaltic plug of Lower Carboniferous age, Dunglass, Strathblane.
D 109	Trap features on the south scarp of the Campsie Fells and the Lower Carboniferous vents of Dumgoyne and Dumfoyne.
D 110	Gorge cut in Upper Devonian sandstones, Finnich Glen, near Killearn.
D 111	Mine adits in Calmy Limestone (Upper Limestone Formation), Darnley Quarry, near Barrhead.

D 112-3	General view of Darnley Quarry showing Calmy Limestone and overlying mudstones (Upper Limestone Formation), near Barrhead.
D 114	General view of Boyleston Quarry showing face of lavas (Clyde Plateau Volcanic Formation), near Barrhead.
D 115	Crag and tail, Meikle Caldon. Crag is formed by Lower Carboniferous vent, tail is orientated to the east-south-east; Stockie Muir.
D 116	Crag and tail, Little Caldon. Crag is formed by Lower Carboniferous vent, tail is orientated to the east-south-east; Stockie Muir.
D 117-8	General view across Stockie Muir to the Loch Lomond area.
D 119-20	Dip and scarp features in the Kilpatrick Hills, near Auchengillan, formed by lavas of the Clyde Plateau Volcanic Formation.
D 121	Trap features of the south scarp of the Campsie Fells and the Lower Carboniferous vents of Dumgoyne and Dumfoyne.
D 122	The Whangie, landslipped mass of basalt from Lower Carboniferous intrusion, showing deep, open back fissure, Auchineden Hill.
D 123-6	Wilderness Till Formation on sand and gravel of Cadder Formation, Wilderness Sandpit, near Bishopbriggs.
D 128	General view of Wilderness Till Formation on sand and gravel of Cadder Formation, Wilderness Sandpit, near Bishopbriggs.
D 129	Wilderness Till Formation on sand and gravel of Cadder Formation, Wilderness Sandpit, near Bishopbriggs.
D 836-8	General view of north scarp of the Campsie Fells showing line of Lower Carboniferous vents.
D 1472	Fossil Grove, natural casts of lower rooted portions of *Lepidodendron* trees (Upper Carboniferous).
D 1535-7	Fossil Grove, natural casts of lower rooted portions of *Lepidodendron* trees (Upper Carboniferous).
D 1851-4	Coarse agglomerate in funnel-shaped neck of vent cutting fine ash, Craigangowan Quarry, Milngavie.
D 1855-6	Inclined columnar jointing in Lower Carboniferous basaltic dyke, Craigangowan Quarry, Milngavie.
D 1857	Columnar jointing in ponded lava of the Clyde Plateau Volcanic Formation, Pillar Crag, near Strathblane.
D 1858	Deep fissure at back of landslip in Lower Carboniferous Intrusion, The Whangie, Auchineden Hill.
D 1859	Discordant relationship of intrusion cut by The Whangie and lavas of the Clyde Plateau Volcanic Formation, Auchineden Hill.
D 1865	Waterfall formed by lava resting on deeply undercut bole and scoria (Clyde Plateau Volcanic Formation), Campsie Glen, Lennoxtown.
D 1870	General view of wasting peat, with plugged conduit (Lower Carboniferous) in background, Garloch Hill, near Killearn.
D 1871	Columnar jointed basaltic intrusion of Lower Carboniferous age, Canny Tops, near Killearn.
D 1872-3	Coarse, unsorted agglomerate, close to or in volcanic vent, the Garlochs, near Killearn.
D 1874	Cross-stratified sandstone of the Clyde Sandstone Formation, the Garlochs, near Killearn.
D 1875	Alternating beds of mudstone and cementstone of the Ballagan Formation, Garloch Well, near Killearn.
D 1876	Cross-stratified pebbly sandstone of the Kinnesswood Formation, Duke's Quarry, near Killearn.
D 1877	Flaggy, microfaulted sandstone of the Kinnesswood Formation, Maucher Glen, near Killearn.
D 1878-9	Immature cornstones and carbonate concretions in sandstones of the Kinnesswood Formation, Maucher Glen, near Killearn.
D 1880	Cross-stratified pebbly sandstone of the Kinnesswood Formation, Maucher Glen, near Killearn.
D 1881-2	Alternating beds of mudstone and cementstone of the Ballagan Formation overlain by sandstone of the Clyde Sandstone Formation, Little Corrie.
D 1883-4	Alternating beds of mudstone and cementstone of the Ballagan Formation overlain by sandstone of the Clyde Sandstone Formation, Ballagan Burn, Strathblane.
D 1885	Alternating beds of mudstone and cementstone of the Ballagan Formation, Little Corrie, near Killearn.
D 2343, 2349	Fossil Grove, natural casts of lower rooted portions of *Lepidodendron* trees, Victoria Park, Glasgow.
D 2995	South scarp of Campsie Fells showing trap features formed by lavas of the Clyde Plateau Volcanic Formation and the Lower Carboniferous vents, Dumfoyne and Dumgoyne.
D 3314	Lang Crags, showing trap features formed by lavas of the Clyde Plateau Volcanic Formation.
D 3315	Columnar jointing in minor basaltic intrusion, near Bowling.
D 3316	Lavas of the Clyde Plateau Volcanic Formation passing into ash, Kilmannon Reservoir.
D 3317	Quartz-conglomerate of the Lawmuir Formation with trough-cross-bedded sandstone lenses, Douglas Muir Quarry.
D 3318	Volcanic detritus of the Kirkwood Formation overlain by quartz-conglomerate, Douglas Muir Quarry.
D 3320	Lobe of one lava flow against two adjacent flows (Clyde Plateau Volcanic Formation), near Old Kilpatrick.
D 3321	Large masses of scoriaceous material probably infilling lava tunnels (Clyde Plateau Volcanic Formation), near Old Kilpatrick.
D 3363-4	Alternating beds of mudstone and cementstone of the Ballagan Formation, Overton Burn.
D 3366	Cross-bedded sandstone of the Clyde Sandstone Formation, Overton Burn.
D 3367-8	Lower Carboniferous vent with large block of Ballagan Formation strata in agglomerate, Katythirsty Well, Quinloch Muir.
D 3369	Stoop and room workings in the Baldernock Limestone (Lawmuir Formation), Linn of Baldernock.
D 3370	Trap features formed by lavas of the Clyde Plateau Volcanic Formation, Lang Craigs.
D 3371	Mudstone of the Lower Limestone Formation quarried for brick manufacture, Blairskaith Quarry, near Milngavie.

APPENDIX 4

D 3372   Scarp formed by lavas of the Clyde Plateau Volcanic Formation with extensive landslip at its base, Lang Craigs.

D 3373   Purple sandstones with thin intercalations of mudstones, of Lower Devonian age, Pots of Gartness.

D 3374   Series of Lower Carboniferous vents at western end of Kilpatrick Hills.

D 3375   Waterfall formed by lava on deeply undercut coal and seatrock (Clyde Plateau Volcanic Formation), Auchentorlie Glen, Bowling.

D 3376   Trap features formed by lavas of the Clyde Plateau Volcanic Formation, Greenland Reservoir.

D 3377-8   Partly silicified mature cornstone in the Kinnesswood Formation, Roughting Burn, near Dumbarton.

D 3379-81   Proximal facies lavas and lapilli tuff of the Clyde Plateau Volcanic Formation, Kilmannon Reservoir.

D 3382   Trap features formed by lavas of the Clyde Plateau Volcanic Formation, Auchineden Hill.

D 3383-4   Crag and tail, Duncryne Hill. Crag is formed by Lower Carboniferous vent and tail is composed of Gartochan Till Formation.

D 3385   House showing different use made of local Upper and Lower Devonian sandstones, Gartocharn.

D 3390-1   Waterfall formed by lava flow lying on the basal volcaniclastic sedimentary rocks of the Clyde Plateau Volcanic Formation, Craigie Linn, Glen Park, Paisley.

D 3394   Lower Carboniferous intrusions of Duncolm and Middle Duncolm, Kilpatrick Hills.

D 3518   Lower Carboniferous vent, Dumbuck, Bowling.

D 3765-6   Upper Devonian aeolian sandstone showing alternations of cross-laminated (dune facies) and parallel laminae (interdune faces), Finnich Glen, near Killearn.

D 4978   The University of Glasgow on drumlin of Wilderness Till Formation.

D 4979 A & B   The University of Glasgow on drumlin of Wilderness Till Formation.

D 4980 A & B   Glasgow Cathedral and the Necropolis. The Necropolis is built on a crag and tail, the crag being formed by an olivine-dolerite sill.

D 4981   Lava scarps of the Clyde Plateau Volcanic Formation on the Hill of Dun and Kilpatrick Braes, framed by Erskine Bridge.

APPENDIX 5

Index of Carboniferous and Devonian taxa

Page number in italics indicates plate.

Actinopteria regularis (Etheridge jun, 1873) [bivalve]   *44*
algae   45
Alitaria spp. [brachiopod]   43
Anthraconaia cymbula (Wright, 1929) [bivalve]   46
A. librata (Wright, 1929)   46
A. polita (Trueman, 1929)   46
Anthracosia aquilina Trueman and Weir, 1951 *non* Sowerby, 1840 [bivalve]   46
A. atra (Trueman, 1929)   46
A. regularis (Trueman, 1929)   46
Anthracosphaerium cf. *trumani* Leitch, 1947 [bivalve]   46
Antiquatonia costata (J de C Sowerby, 1827) [brachiopod]   45
A. hindi (Muir-Wood, 1928)   43
A. insculpta (Muir-Wood, 1928)   43
A. sulcata (J Sowerby, 1822)   43
Avonia youngiana (Davidson, 1860) [brachiopod]   43

Beecheria hastata (J de C Sowerby, 1824) [brachiopod]   43
Bothriolepis [fish]   9
Buxtonia sp. [brachiopod]   43–44

Caneyella membranacea (McCoy, 1844)   *44*
Carbonicola oslancis Wright, 1929 [bivalve]   46
C. pseudorobusta Trueman, 1929   40
Carbonita sp. [arthropod]   29
cephalopod   37
Composita ambigua (J Sowerby, 1822) [brachiopod]   43
conodonts   46–47
coral   27, 30, 38, 39
Cravenoceras [goniatite]   8
'*C.*' *gairense* Currie, 1954   45
C. leion Bisat, 1930   43
C. scoticum Currie, 1954   43
crinoid   31, 38
Crurithyris urii (Fleming, 1828) [brachiopod]   31, 43
Curvirimula sp. [bivalve]   31, 37, 39–40, 45–46

Dimorphoceras marioni Moore, 1939 [goniatite]   43
Diphyphyllum sp. [coral]   30

Echinoconchus elegans (McCoy, 1855) [brachiopod]   43
E. punctatus (Martin, 1809)   43, *44*
Edmondia punctatella (Jones, 1865) [bivalve]   45–46

Eomarginifera spp. [brachiopod]   43
'*Estheria*' sp. [conchostracan]   14, 43, 46
Euchondria neilsoni Wilson, 1966 [bivalve]   31
Euestheria sp. [conchostracan]   41, 46
Eumorphoceras ferrimontanum Yates, 1962 [goniatite]   45
E. grassingtonense Dunham and Stubblefield, 1945   45
Euphemites ardenensis (Weir, 1931) [gastropod]   46
E. sp.   45

fish   21, 43
foraminifera   46–47, 70, 78

Gastrioceras subcrenatum (C Schmidt, 1924) [goniatite]   8
gastropod   31, 37, 45
Geisina arcuata Bean, 1836 [ostracod]   40, 46
Glabrocingulum atomarium (Phillips, 1836) [gastropod]   31
goniatite   31, 39, 43, 45

Holinella sp. [arthropod]   46
Holoptychius [fish]   9

Latiproductus cf. *latissimus* (J Sowerby, 1822) [brachiopod]   43, 45
L. sp.   43
Leaia sp. [conchostracan]   40, 46
Lepidodendron [plant]   45
Leptagonia smithi Brand, 1972 [brachiopod]   *44*
Lingula mytilloides J Sowerby, 1812 [brachiopod]   41, 46
L. squamiformis Phillips, 1836   45
L. sp.   6, 30, 37, 39, 43–44
Linoproductus cf. *concinniformis* (Paeckelmann, 1931)   *44*

Meekospira sp. [gastropod]   45
Myalina? [bivalve]   46

Naiadites obliquus Dix and Trueman, 1932 [bivalve]   46
N. sp.   37, 39, 45–46

Orbiculoidea sp. [brachiopod]   43, 45–46
ostracod   14, 30–31, 37, 43, 45–46

Palaeoneilo luciniformis (Phillips, 1836) [bivalve]   45
P. mansoni Wilson, 1967   45
Paracarbonicola sp. [bivalve]   37, 45

Pernopecten fragilis Wilson, 1966 [bivalve]   31
Phricodothyris lineata (J Sowerby, 1822) [brachiopod]   43
Planolites ophthalmoides Jessen, 1949 [trace fossil]   46
P. aff. *ophthalmoides*   46
Pleuropugnoides sp. [brachiopod]   43
polyzoa   38
Posidonia becheri Bronn, 1828 [bivalve]   43
P. corrugata Etheridge jun, 1873   37, 45
P. c. gigantea Yates, 1962   31, *44*
P. membranacea (McCoy, 1844)   43
Productus concinnus J Sowerby, 1821 [brachiopod]   43
P. sp.   27, 44
Pugilis pugilis (Phillips, 1836) [brachiopod]   43
Pugnax cf. *pugnus* (Martin, 1809) [brachiopod]   *44*, 46

Rayonnoceras windmorense Selwyn-Turner, 1951 [nautiloid]   44
Retispira sp. [gastropod]   45
Rhipidomella michelini Leveillé, 1835 [brachiopod]   43
Rugosochonetes sp. [brachiopod]   43

Schizophoria resupinata (Martin, 1809) [brachiopod]   43
Serpuloides carbonarius (McCoy, 1844) [annelid]   45–46
Sinuatella cf. *sinuata* (de Koninck, 1851) [brachiopod]   46
Siphonodendron sp. [coral]   30
Spirifer bisulcatus J de C Sowerby, 1825 [brachiopod]   43
S. crassus (Martin, 1809)   43
S. crassus/striatus (Martin, 1809) group   43
Spirorbis sp. [annelid]   14, 43
Straparollus carbonarius (J de C Sowerby, 1844) [gastropod]   31, 45
Streblopteria ornata (Etheridge jun, 1873) [bivalve]   45
Sudeticeras splendens (Bisat, 1928) [goniatite]   43

Teichichnus sp. [trace fossil]   34
Tornquistia polita (McCoy, 1852) [brachiopod]   43
T. youngi Wilson, 1966   31
trilobite   38
Tumulites pseudobilinguis (Bisat, 1922) [goniatite]   45

APPENDIX 6

Distribution and index of Quaternary taxa

Page number included where mentioned in text; in italics indicates plate.

	FORMATION							FORMATION					
	Ba	In	Py	Ln	Er	Go		Ba	In	Py	Ln	Er	Go
MARINE ALGAE							*Lagena semilineata* Wright, 1886			*			
Lithothamnion sp.						*	*Lagena striata* (d'Orbigny, 1839)			*			
ANNELIDA							*Lagena substriata* Williamson, 1848				*		
Pomatoceras sp.						*	*Lagena sulcata* (Walker and Jacob, 1798)			*	*		
Spirorbis sp.						*	*Lagena vikensis* Hessland, 1943				*		
FORAMINIFERIDA							*Lenticulina* cf. *thalmanni* (Hessland, 1943)				*		
Ammonia batavus (Hofker, 1951)	*						*Massilina secans* (d'Orbigny, 1826)				*		
Ammonia beccarii (Linné, 1758)			*				*Milliamina fusca* (Brady, 1970)			*	*		
Bolivina pseudoplicata Heron-Allen and Earland, 1930				*			*Milliolinella subrotunda* (Montagu, 1803)	*			*		
Bolivina cf. *pseudopunctata* Höglund, 1947				*			*Milliolinella* sp.	*			*		
Bolivina sp.		*		*			*Neogloboquadrina pachyderma* (Ehrenberg, 1861)			*	*		
Bucella frigida (Cushman, 1922)	*			*			*Nonion labradoricum* (Dawson, 1860)			*			
Bulimina gibba Fornasini, 1960			*				*Nonionella turgida* (Williamson, 1858)				*		
Bulimina marginata d'Orbigny, 1826	*						*Patellina corrugata* (Williamson, 1858)	*					
Buliminella borealis Haynes, 1973			*				*Pseudopolymorphina novangliae* (Cushman, 1923)						
Cassidulina reniforme Nörvang, 1945	*	*		*			*Pyrgo williamsoni* (Sylvestri, 1923)	*			*		
Cibicides lobatulus (Walker and Jacob, 1798)	*	*		*			*Quinqueloculina* cf. *cliarensis* Heron-Allen and Earland, 1930				*		
Dentalina cf. *drammenensis* Fehling-Hanssen, 1964				*			*Quinqueloculina lata* Terquem, 1876			*	*		
Elphidium albiumbilicatum (Weiss, 1954)		*		*			*Quinqueloculina seminulum* (Linné, 1767)	*		*	*		
Elphidium asklundi Brotzen, 1943	*						*Rosalina williamsoni* (Chapman and Parr, 1932)				*		
Elphidium bartletti Cushman, 1933	*			*			*Textularia* sp.				*		
Elphidium clavatum Cushman, 1930 71		*	*	*			*Trifarina fluens* (Todd, 1947)	*					
Elphidium clavatum/lidoense Cushman				*			*Virgulina loeblichi* Fehling-Hanssen, 1954			*	*		
Elphidium exoticum Haynes, 1973				*			polymorphinids	*			*		
Elphidium cf. *exoticum* Haynes				*			GASTROPODA						
Elphidium cf. *gerthi* van Voorthuysen, 1957						*	*Acmaea* sp.						*
Elphidium incertum (Williamson, 1858)			*	*			*Alvania punctura* (Montagu, 1803)						*
Elphidium subarcticum Cushman, 1944		*		*	*		*Bittium reticulatum* (da Costa, 1778)						*
Elphidium williamsoni Haynes, 1973			*	*			*Cylichna cylindracea* (Pennant, 1777)			*	*		*
Fissurina laevigata Reuss, 1850				*			*Eulima*?						*
Fissurina lucida (Williamson, 1848)		*		*			*Gibbula cineraria* (Linné, 1758)						*
Fissurina marginata (Walker and Boys, 1784)	*						*Lacuna vincta* (Montagu, 1803)			*	*		*
Fissurina orbignyana Sequenza, 1862			*	*			*Littorina littorea* (Linné, 1758)						*
Guttulina?				*			*Littorina rudis* (Maton, 1797)				*		
Haynesina germanica (Ehrenberg, 1840)		*		*			*Littorina saxatilis* (Olivi, 1792)						*
Haynesina orbiculare (Brady, 1881)	*			*			*Littorina* spp.				*		
Lagena clavata (d'Orbigny, 1846)				*			*Lunatia alderi* (Forbes, 1838)						*
Lagena distoma Parker and Jones, 1864	*			*			*Lunatia pallida* (Broderip and Sowerby, 1829)			*	*		
Lagena gracilis Williamson, 1848			*	*			*Lunatia* sp.			*	*		
Lagena laevis (Montagu, 1803)	*		*	*			*Nassarius incrassatus* (Ström, 1768)						*
							Oenopota trevelliana (Turton, 1834)				*		
							Oenopota turricula (Montagu, 1803) 79, 81–82			*	*		*
							Oenopota sp.				*		

FORMATION

	Ba	In	Py	Ln	Er	Go
Onoba semicostata (Montagu, 1803)			*			*
Philbertia linearis (Montagu, 1803)						*
Pusillina sarsi Lovén, 1846						*
Retusa obtusa (Montagu, 1803)			*			
Retusa umbilicata? (Montagu, 1803)						*
Retusa sp.			*			
Rissoa parva (da Costa, 1778)			*			
Rissoa parva interrupta (Adams, 1798)						*
Rissoa rufilabrum Alder, 1844						*
Skeneopsis planorbis (Fabricius, 1780)			*	*		
Tectonatica clausa (Broderip and Sowerby, 1829)						*
Turbonilla elegantissima (Montagu, 1803)						*
Turritella communis Risso, 1826						*
BIVALVIA						
Acanthocardia echinata (Linné, 1758)						*
Arca pectunculoides Scacchi, 1836						*
Arctica islandica (Linné, 1767) 70		*	*			*
Chlamys opercularis (Linné, 1758)						*
Cochlodesma?						*
Dosinia exoleta (Linné, 1758)						*
Ensis siliqua (Linné, 1758)						*
Fabulina fabula (Gronovius, 1781)						*
Gari fervensis (Gmelin, 1791)						*
Heteranomia squamula (Linné, 1758)						*
Hiatella arctica (Linné, 1767) 56, 57		*		*		*
Laevicardium sp.						*
Lucinoma?						*
Macoma balthica (Linné, 1758)						*
Macoma calcarea (Gmelin, 1790) 56, 57, 71				*		
Montacuta ferruginosa (Montagu, 1803)						*
Musculus discors (Linné, 1758)			*	*		
Musculus sp.		*				
Mya truncata Linné, 1758 56, 57, 78			*	*		
Myrthea spinifera (Montagu, 1803)						*
Mytilus edulis Linné, 1758 56, 57, 79			*	*	*	
Nicania montagui (Dillwyn, 1817)				*		
Nucula belloti (Adams, 1856)				*		
Nucula sp.			*	*	*	
Nuculana minuta (Müller, 1776)				*		
Nuculana pernula (Müller, 1776)		*	*			
Nuculana sp.			*	*		
Parvicardium ovale (Sowerby, 1841)				*		
Parvicardium sp.			*	*		
Portlandia arctica (Gray, 1824) 72, 76		*				
Similipecten similis? (Laskey, 1811)						*
Spisula elliptica (Brown, 1827)						*
Spisula subtruncata (Da Costa, 1778)						*
Thyasira flexuosa (Montagu, 1803)						*
Thyasira cf. *gouldi* (Philippi, 1845)				*		
Thyasira sp.			*			
Tridonta elliptica (Brown, 1827)				*		
Tridonta sp.			*			
Venerupis pullastra? (Montagu, 1803)						*
Venus casina Linné, 1758				*		

FORMATION

	Ba	In	Py	Ln	Er	Go
Venus ovata (Pennant, 1777)						*
Venus striatula (Da Costa, 1778)						*
Yoldiella fraterna (Verrill and Bush, 1898)				*	*	
Yoldiella lenticula (Möller, 1842)	*			*	*	
astartacean fragments				*		
cardiacean indet.				*		
mytilacean fragments	*			*		
OSTRACODA						
Acanthocythereis dunelmensis (Norman, 1865)	*			*	*	
Bythocythere constricta Sars, 1866				*		
Celtia quadridentata (Baird, 1850)				*		
Cluthia cluthae (Brady, Crosskey and Robertson, 1874)				*	*	
Cyprideis torosa (Jones, 1850)						
Cythere lutea (Müller, 1785)				*		
Cytherois fischeri Sars, 1866						
Cytheropteron angulatum Brady and Robertson, 1872						
Cytheropteron biconvexa Whatley and Masson, 1979	*					
Cytheropteron latissimum (Norman, 1865)				*		
Cytheropteron nodosum Brady, 1868	*			*		
Cytheropteron pararcticum Whatley and Masson, 1979				*	*	
Cytheropteron pseudomontrosiense (Whatley and Masson, 1979)				*		
Cytheropteron pyramidale Brady, 1868				*		
Cytheropteron subcircinatum Sars, 1866				*		
Cytheropteron cf. *subcircinatum* Sars, 1866				*		
Elofsonella concinna (Jones, 1857)	*			*		
Eucythere declivis (Norman, 1865)				*		
Eucytheridea macrolaminata (Elofson, 1939)				*		
Finmarchinella angulata (Sars, 1866)				*		
Hemicytherura cf. *clathrata* (Sars, 1866)				*		
Heterocythereis albomaculata (Baird, 1838)				*		
Hirschmannia viridis (O F Müller, 1785)						
Jonesia acuminata (Sars, 1866)				*		
Leptocythere castanea (Sars, 1866)				*		
Palmenella limicola (Norman, 1865)				*		
Palmoconcha laevata (Norman, 1865)				*		
Paradoxostoma variabile (Baird, 1835)						
Polycope orbiculare (Sars, 1865)				*		
Pontocypris mytilloides (Norman, 1862)				*		
Pontocythere elongata Brady, 1868						
Robertsonites tuberculata (Sars, 1866)				*		
Roundstonia globulifera (Brady, 1868)				*		
Sarsicytheridea bradii (Norman, 1865)	*			*		
Sarsicytheridea punctillata (Brady, 1865)	*			*		
Sclerochilus contortus Norman, 1861						
Semicytherura nigrescens (Baird, 1838)	*			*		
Semicytherura undata (Sars, 1866)				*		
Semicytherura sp.				*		
Xestoleberis depressa Sars, 1866				*		

	FORMATION					
	Ba	In	Py	Ln	Er	Go
CRUSTACEA						
Balanus sp.	*		*	*		*
Elminius modestus Darwin, 1854						*
ECHINODERMATA						
echinoid fragments			*			*

Ba = Balloch Formation
In = Inverleven Formation
Py = Paisley Formation
Ln = Linwood Formation
Er = Erskine Formation
Go = Gourock Formation

Index of other Quaternary taxa mentioned in the text

Alnus 78–79
Balanus balanus Linné, 1767 [crustacean]   56, 57
Boreotrophon clathratus (Linné, 1767) [gastropod]   56, 57, 74
Buccinum undatum Linné, 1758 [gastropod]   56, 57
Calluna 82
Chlamys islandica (Müller, 1776) [bivalve]   56, 57, 71, 74, 76
Corylus 78–79
Ericales 82
Modiolus modiolus (Linné, 1758) [bivalve]   70
Mytilus 54
Sphagnum 81
Plantago 72
Rumex 72
Tectonatica clausa (Broderip and Sowerby, 1829) [gastropod]   56, 57, 74
Tridonta elliptica (Brown, 1827)   56, 57
T. montagui (Dillwyn, 1817) [bivalve]   56, 57
Ulmus 78–79
Unio margaritifera (Linné, 1758) [bivalve]   81

Index of other Quaternary fossils mentioned in the text

Alder   78
Balanid   78
Barnacle   76
Beetle   78
Deer   63
Diatom   81
Dinoflagellate cyst   78–79
Hazel   78
Oak   81

INDEX

Aegiranum (Skipsey's) Marine Band 6, 7, 40, 42, 46
aeolian facies and sandstones 4, 5, 50, 86
aeromagnetic surveys 88, 91
agglomerate 15, 18
Airdrie Virtuewell Coal 40, 46, 83
airfall tuff 21
Aitkenhead No. 4 Borehole 48
albitisation 17
Aldessan Burn 18, 19
Alexandria Parade 63–64
alkali and basic basalts 6, 17–19, 22, 23, 24, 26, 50
alkali dolerite sills and dykes 48–49, 84, 93, 95
Allander Water 58
alluvium 82
Alportian Stage 7
Alum Shale 30, 83, 84
Alvain Burn 18–19
anhydrite 14
ankaramite 21, 22, 26
Arden Coal 39
Arden Limestone *see* Calmy Limestone
Ardrossan Canal 54
Ardyne 70
arid and semi-arid deposition 6, 9
Arkleston 48, 54
Arlehaven 22
Arnsbergian Stage 7, 37, 45
Arthurlie Coal 39
Arundian Stage 7
Asbian Stage 7, 27, 47
Ashfield Coking Coal 37
Ashfield Rider Coal 34
Auchentorlie 22
Auchinback Coal 39
Auchincarrock Quarry 84
Auchineden Hill 14, 77
Auchineden Lavas 21, 22, 24, 25, 26
Auchingyle Farm water borehole 85

Baillieston Till Formation 58, 59, 60, 61
baked limestone 29
Baldernock (White) Limestone 27, 29, 30, 83
Baldernock Mill 29
Baldow Glen 29
Balglass 76
Balgrochan Beds 27, 29
Baljaffray 31
Ballagan Formation 7–8, 13–14, 15, 43, 46
Ballagan Glen 13, 15
Ballikinrain Muir 13
Balloch Formation 58
Balmaha car park water borehole 85

Balmore 27, 29, 34, 37, 86
Balmore Fault 53
Balmore Marine Band 27, 43
Balmore No. 2 Borehole 31
Balmuildy 87
Banner Road 58
Banton Rider Coal 37
Bardowie 53
Bargaran Burn 80
Barhill Plantation 63
Barnhill Borehole 14, 43
Barochan Moss 79, 80, 85
Barr Hill 84
Barraston Burn 29
Barrhead 34, 37, 39, 84
Barrhead Fault 53
  see also Dusk Water-Barrhead Fault
Barrhead Grit 39
Barrhead Hills *see* Beith-Barrhead Hills and succession
Barshaw 48
basalt 6, 17–19, 22, 23, 24, 25, 26, 50, 77
basement fractures, deep-seated 53
basin peat 55
Batchie Ironstone 37
Bearsden 57, 63, 86
Bearsden low 90, 91
Bearsden Station 61, 63
Beith-Barrhead Hills and succession 15, 17, 20, 23–24, 49, 52, 90, 93, 95
Bellshill Formation 58, 64, 66, 84
Bellside Coal 40
benmoreite 26
Berryhills Limestone 34
'big-feldspar' basalt 24
biostratigraphy of Carboniferous 8
bioturbation 34, 39
Birdstone 66
Birny Hills 22
Bishopbriggs 38, 39, 60–61, 63, 82, 84
Bishopbriggs No. 2 Sandpit 66
Bishopbriggs Sandstone 36, 37, 39, 84
Bishopton 27, 48
Black Cart Water 57, 82
Black Craig Lavas 18
Black Loch 21, 22, 23
Black Metals 34, 36
Black Metals Marine Band 37, 44
blackband ironstone 34, 36–37, 45, 83
Blackbyre Limestone 30, 43, 83
Blackhall Limestone 31, 43, 48, 49, 83
Blackhill Brickworks 87
Blackhill Colliery 39
Blackhill Syncline 52–53
Blackloch Vent 23
Blairdardie No. 4 Pit 58
Blairquhomrie Muir 84
Blairskaith 33, 84
Blane Water and valley 54, 55, 57, 82, 84–86
Blane Valley Fault 53
Blane Water Formation 58, 72, 74, 75, 84–86
  see also Strath Blane
Blanefield 10, 13, 50, 76
blanket bog 81

Blantyreferme 64, 66
Blythswood Fault 53, 90, 91
Boclair No. 6 Borehole 58
bole 15, 23
Bolsovian (Westphalian) Stage 7, 8, 39, 42, 43, 51, 90
Bonny Water 54, 64
Bothwell Bridge Marine Band 42, 46
Bouguer gravity surveys 88, 93
boulder clay *see* till
Bowling 21, 22, 50
Bowling Centre 22
Bowling Vent 23
bowlingite 24
Boylestone 24, 84
braided streams 14
Braidfield Pit 57
breccias and brecciated zones 14, 15
Brediland 30, 83
brick clay 61, 84, 87
brickclay manufacture 84
Bridgeton 57
Bridgeton Borehole 62, 66, 69, 71, 81
Bridgeton Formation 58, 66, 67, 70
Brigantian Stage 7, 27, 31, 43, 47
Broad Street 40
Broomhill Boreholes 63
Broomhill Park No. 2 Borehole 61
Broomhill Formation 58, 59, 61, 63, 67
Broomhouse Formation 58, 64, 66
Brown Hill 21
Brown Hill Vent 22
Brownside 30, 83
Brownside Braes 23
Buchanan Formation 58, 79
Buchley 87
Buchley Till Member 61
buried ice masses 66, 67
Burnbrae 21
Burnbrae Reservoir 17
Burncrooks 21, 22
Burncrooks Pyroclastic Member 21, 22, 25, 26
Burnside Vent 22, 23

Cadder 39, 53, 63, 66, 85
Cadder Formation 58, 60, 61, 63, 85
Cadder No. 1 Borehole 39
Cadder No. 6 Borehole 37, 38, 39
Cadder No. 16 Borehole 60, 66
Cadder No. 17 Pit 83
Cadgers Loan Sandstone 39
Calciferous Sandstone Measures 6
Caledonia Fireclay Pit 29, 30, 84
Caledonoid trend 50, 51, 53
California Ironstone 36
Calmy Limestone 37, 38, 39, 44, 45–46, 83, 84
Cambuslang 57
Cambuslang Marble 41, 46
Cameron Burn 74
Cameron Muir 72, 85
Campsie Block 17, 18, 26, 27
Campsie Clayband Ironstone 33
Campsie Dyke 48
Campsie Fault 17, 19, 24, 90, 93

Campsie Fells 13, 14, 15, 16, 17, 19, 20, 24, 25, 50, 55, 57, 64, 76, 77, 81, 85, 90
Campsie Glen 15
Campsie Lavas 18, 24, 25
Capellie Farm 24
Carbeth House 74
Carbeth Lavas 21, 22, 25, 26
carbonate concretions 12, 13, 14
carbonation 17
Carboniferous 7–49
 chronostratigraphy and biostratigraphy 8
 classification 6–8
 faults 17, 19, 24, 27, 30, 40, 42
 geophysical investigations 90, 91, 92–93, 95
 intrusive rocks in 48–49
Cardonald 48
Carmyle 71
Carnock Burn 4, 72
Carron, River 54
Cart Harbour 71
Castlecary Limestone 7, 34, 37, 39
Castlehead Lower Coal 29, 30, 83
Castlehead Pit 29
Castlehead Upper Coal 29, 30
Castlemilk Fault 90
Castlemilk House 48
Castlemilk West Fault 53
Cathcart Castle 48
Cathcart high 91
Cathcart sills 48, 49
Cathkin Braes 15, 17, 24
Catter Burn 3, 72
Catythirsty Vent 16, 21
Cawder Cuilt 39
Cawder Cuilt No. 1 Borehole 39
cementstone 13, 14
Chadian Stage 7
channel deposits 14, 81
channel-fill sequences 9, 12, 13
channel sandstone 41
Chapelgreen Coal 39
chlorite/chloritisation 17, 49
Chokierian Stage 7
chronostratigraphy of Carboniferous 8
Clachan of Campsie 19, 20, 22, 63
Clachan of Campsie Vent 23
Clachie Burn 18
Clackmannan Group (Dinantian part) 6, 7, 31–33
Clackmannan Group (Silesian part) 6, 7, 34–40
Clarkston Fault 90, 93
clay
 brick 61, 84, 87
 fireclay 29, 30, 39, 84
 resources 84, 86
 and silt, Quaternary 54, 55, 58–84 *passim*
clayband ironstone 31, 33, 36, 45, 83
cliffline 81
Clippens Peat Formation 58, 70, 79, 81, 82
Clyde Beds 54

Clyde Ironworks 46
Clyde Plateau Volcanic Formation 7, 13, 15–27, 49, 50, 52, 53, 84, 85, 86, 88, 90–91, 92, 93
 classification 17
Clyde Road Tunnel 64, 82
Clyde Sandstone Formation 7, 14–15
Clyde valley 55, 58, 64, 71, 79–85, 86
Clydebank 27, 76, 85, 88
Clydebank Cemetery 64, 71
CM miospores zone 8
coal
 coking 30, 34, 83
 -cyclic sequence 34, 37, 39, 40
 economic geology 83
 -forming vegetation 37, 40, 42
 gas 34, 36, 83, 86–87
 stoop and room workings 29, 86
 swamp 30, 40
 working hazards 86–7
Coal Measures 6, 7, 40–42, 44, 45, 46, 51, 52, 53, 83
 Lower 6–8, 29, 40–41, 46, 51, 83
 Middle 6–7, 40–42, 46, 83
 Upper 6–7, 29, 41–42, 46, 51–53, 83–84
coastal plain 14
Coatbridge Balmoral Coal 40
Coatbridge Musselband 41
Cochno Lavas 19, 21, 22–23, 24, 25
Cochno Loch 22
Cochno Reservoir 17
coking coal 30, 34, 83
Colston Road Borehole 34, 48
Communis Chronozone 40
concretion 9, 13
conduit 16
conglomerate resources 84
contaminated land 87
continental rift environment 26
copper 14, 24
cornstone 4, 9–10, 12–13, 14, 83
Corrie of Balglass 57
Corseford Limeworks 44
Courcevan Stage 7, 46
Cowlairs Sandstone 34
crag-and-tail landform 57, 63
Craigallian Loch 22
Craigangowan Quarry 16
Craigend 84
Craigend Castle 23
Craigengaun 84
Craigenglen Beds 27, 29
Craigentimpin Lavas 18–19, 24
Craigentimpin Quarry 19
Craighead Coal 30
Craiglockart Basalt/lava 17, 18, 22, 23, 24, 25, 26
Craigmaddie Muir 27
Craigmaddie Sandstone 27, 29
Craigton 21, 22
Craigton Vent 23
Croftamie 3, 4, 5, 74, 84
Croftamie high 89, 93
Crookston No. 1 Borehole 30
Cullen's Pit 57
Cumbernauld 91

'cumulate' texture 23
cuprite 24
cyclic sedimentation 50
cyclothem 6

Dale Street 81
Dalmeny Basalt/lava 17, 18, 22, 23, 24, 25
Dalmuir 54
Dalnair 74
Dalreoch Quarry 4
Danes Drive, Scotstoun 76
Darnley 39, 83–84
Darnley Basin 34, 37–38, 52
Darnley Fireclay 39, 84
Darnley No. 3 Borehole 44
Darnley No. 4 Borehole 38
Darnley Quarry 44
Davieland 44
dead ice 64, 66, 67, 69
decaying glacial ice 64
Dechmont Fault 40, 42, 50, 53, 91
deformation till 61, 64
degassing cavities 76
delta/deltaic deposits
 Lower Carboniferous 33
 Quaternary 54, 61, 66, 71
 Upper Carboniferous 37, 39, 40, 42
deposition
 cyclic 50
 depositional basin 33
 Devonian 3, 4–5, 6, 50
 Lower Carboniferous 6, 7, 8, 9, 14, 27, 29–31, 33
 Quaternary 54–55
  Dimlington 63
  Flandrian 78–82
  Loch Lomond Stadial 76–78
  pre-Dimlington 57–63
  Windermere Interstadial 64–72
 Upper Carboniferous 6, 40–41, 42, 45
  Clackmannan Group 34, 36, 37, 38–39
dessication cracks 14, 15
Devensian topography and palaeontology 55–57
Devil's Craig Dam 23
Devonian 3–6
 deposition 3, 4–5, 6, 50
 geophysical investigations 88–89, 90, 91, 92, 93, 95
 mid-Devonian uplift 50
dewatering 86
diamicton
 Dimlington Stadial 63
 Loch Lomond Stadial 72, 74, 78
 pre-Dimlington Stadial 58, 61
 Windermere Interstadial 64, 69, 70
Dimlington Stadial ice sheet 54, 63–64
 advance of 60, 61
 deglaciation at end of 65
 deposition 63, 77
 erosion 57
 glaciotectonics 63–64
 lithostratigraphy 58

Dinantian *see* Lower Carboniferous
dolomite 9, 14
dolomitic limestone 29, 38
Dougalston Loch 70
Doughnot Hill 21
Doughnot Hill Vent 22
Douglas Muir 23, 27, 84
Douglas Muir Quartz-Conglomerate Member 27, 29, 50, 84
drift resources 84–85, 86
driftwood 81
dropstones 61, 69, 70, 76, 79
Drumbeg Formation 58, 72, 74
Drumbeg Quarry 72, 74
Drumchapel Pit 58–59
Drumcross Farm 78
Drumgray Coal 40, 46
Drumgray Musselband 46
drumlin 57, 63
 inter-drumlin areas 55, 82
Drumpark Marine Band 41
Drumquhassle 72, 74
Drumry 57, 88
Drumry Wood 63
Drymen Pit 72
Duckmantian (Westphalian B) Stage 7, 40, 43
Dumbarton 70
Dumbarton-Fintry line 17, 53
Dumbarton Muir 12, 83
Dumbreck Cloven Coal 37
Drumbeck Sandstone 38, 39
Dumbuckhill 84
Dumfoyne Vent 16, 24
Dumgoyach Brae 14
Dumgoyne Vent 16–17, 24, 79
Duncryne Vent 16
dune facies 4
Dungoil Linear Vent System 18, 19
Dunsapie Basalt/lava 17, 18, 19, 22, 23, 24
Duntocher 29, 31, 83
Dusk Water Fault 24, 95
Dusk Water-Barrhead Fault 91
Dykebar Coal, Upper and Lower 29
Dykebar Limestone 29, 30, 43, 47
Dykebar Marine Band 30
Dykebar Marls 30
dyke 48, 49, 51, 84, 93

East Arlehaven 74
Easter Catter 74
Easter Drumquhassle 74
economic geology 83–87
 drift resources 84–85
 geological hazards 86–87
 groundwater resources 85–86
 solid resources 83–84
Elderslie 30, 48, 83
emerged land surface 37
Endrick Formation 58, 79, 82
Endrick Water and valley 3, 54, 55, 57, 67, 72, 74, 76, 82, 84, 85, 86
erosion
 erosive-based sandstone 14
 fluvial 37, 39, 40
 glacial 54, 57, 63

erratics 72
Erskine Bridge 59, 64, 71
Erskine Bridge Borehole 61, 62, 63, 76
Erskine Formation 58, 79, 81
esker 64
essexite 63
estuarine deposits 55, 81
evaporitic minerals 14

Fairfield Dock 70
Fairy Knowe Quarry 9
Famennian Stage 4, 9
faults 3, 17, 19, 24, 27, 30, 40, 42, 49, 50, 51, 52, 53, 90, 91, 92, 93, 95
Fereneze Hills 24
Fereneze Lavas 23, 24
Ferguslie Fireclay Works and pit 30, 84
Ferguston Hill 61
Fill 82
Fin Glen 15
Fin Glen Lavas 18, 25
Finland Burn 12
Finnich Glen 4, 72
fireclay 29, 30, 39, 84
fissure, water-bearing 86
Flandrian 55, 78–82
flocculation 70
floodplain and flood basin
 Carboniferous 9, 14, 40
 Devonian 3, 5
 Quaternary 55
Fluchter 84
fluviatile/fluvial environments/deposition 14, 51
 Carboniferous 6, 9, 27, 30, 40
 Devonian 3, 5, 50
 Flandrian 82
 erosion 37, 39, 40
 fluviodeltaic environment 50, 51, 61, 64
fold structures 51–53
Forking Burn 18
Fossil Grove 45, 48
Fourteen-Inch Under Coal 34, 44–45
full-pipe sliding bed phase 64
Fullwood Brickfield 57
Fulwood Moss 79, 81, 85
Fynloch 21, 22

Gaidrew 74
Gallangad 12
Gallangad Muir 74
Gallowhill, Paisley 57
Garden Festival, Glasgow 85
Gargunnock Hills 13
Garibaldi Coal 37
Garibaldi Ironstone 36, 83
Garngad 38
Garnieland 30, 83
Garscadden 58
Garscadden Ironstone 36–37, 83
Garscadden Mains 72, 77
Garscube Colliery 83
Gartnavel No. 3 Borehole 48
Gartness Borehole 72–74
Gartness Fault 3, 88–89

Gartocharn 3, 72, 83
Gartocharn Quarry 3
Gartocharn Till Formation 58, 72, 74, 76, 78
Garvald Lavas 24
gas coal 34, 36, 83, 86–87
geliflucted drift 76, 78, 81
geochemistry of Clyde Plateau Volcanic Formation 26–27
geophysical investigations 88–95
 interpretation profile 93–95
 physical properties 92–93
geothermal gradient 92
geothermal potential 92
Giffnock 38, 84
Giffnock Main Coal 36, 39
Giffnock Sandstone 39, 84
glaciofluvial deposits 55, 58, 60–61, 64, 66
glaciolacustrine deposits 55, 58, 61, 63, 64, 66, 74–76
glaciomarine deposits 54–55, 61, 66, 67, 69–72
glaciotectonics 63–64
glaebules 9
Glasgow Main Coal 41
Glasgow Road, Clydebank 76
Glasgow Shale Coal 34, 36, 37
Glasgow Upper Coal 41, 42, 46
Glasgow Upper Marine Band 41, 46
Glazert Water and valley 64, 66, 82, 84
Glen Park, Paisley 23
Glenarbuck 22
Glenboig Limestone 39
Glenboig Marine Band 39
Glenburn 29
Glenburn Borehole 13, 23
Glenburn Reservoir 23, 24
Glenburn Volcanic Detrital Member 23
Glengoyne Distillery 86
Gleniffer Braes 14, 15, 23, 50
Gleniffer Lavas 23
Glenwynd 29
Gonachan Burn 18
Gonachan Glen Linear Vent System 18, 19
Gourock Formation 58, 79, 81
Govan 85
Govan No. 5 Pit 40
 Underground Borehole 38, 39
Govanhill 41, 46, 83
Govanhill No. 27 Borehole 46
gravity investigations 88–91, 93, 94
Greenfieldmuir 23
Greenoakhill 67
Greenside Volcaniclastic Member 21, 22, 25, 43, 47
Greeto Lavas 25
groundwater resources 85–86
Gryfe, River 57, 82
gypsum 14

halite pseudomorphs 14
Hamilton Hill 84
Hamilton Road Route Borehole No. 20 41

114 INDEX

Hampden Park No. 9 Borehole 40
Hampden Park No. 11 Borehole 38
hard rock resources 84
Harelaw Burn 23, 24
Harelaw Reservoir 24
hawaiite 17, 18, 21, 25, 26
Hawthornhill 70
hazards, geological 86–87
heulandite 21, 24
High Craig 84
High Craigton 17
Highland Border Complex 89, 90, 92, 93
Highland Boundary Fault 50, 89, 93
Hillhead 86
Hillhouse Basalt/lava 17, 18
Hirst Coal 39
Holehead Lavas 19, 24, 25
Holkerian Stage 7, 47
Hollybush 83
Hollybush Coal 30
Hollybush Limestone 29, 30, 43, 52, 83
Hosie Sandstone 31
Househill Clayband Ironstone 31, 83
Houston 57
Howwood 27, 31
Howwood Syncline 43
Humph Rider Coal 46
Humphrey, Loch 22, 43, 47
Humphrey Burn 22
Huntershill 36
Huntershill Cement Limestone 37, 38, 39, 45
Hurlet 48–49, 84
Hurlet Borehole 29, 30
Hurlet Coal 27, 29, 30, 49, 57, 83, 84
Hurlet Limestone 6, 7, 27, 30, 31, 43–44, 52, 83–84
Hutchesontown 81
hydraulic fracture 14
hydrogeology 85–86
hydrothermal veins 24

ice
  buried masses 66, 67
  clearance deposits 66–69
  -contact deposits 61, 64, 66
  -dammed lake 59, 72
  dead 64, 66, 67, 69
  decaying 64
  -front 66
  -rafted dropstones 70
  wastage deposits 64
  -wedge casts 61
impoverished fauna 38
Inchinnan 43, 72–73, 77–78
Inchinnan No. 1 Borehole 30
Inchinnan Sewer 57
Index Limestone 7, 34, 37, 38, 39, 45
inter-drumlin areas 55, 82
interdune facies 4
intertidal deposits 81
intrusive rocks
  economic geology 84
  geophysical investigations 90, 91, 92, 93, 95

late and post-Carboniferous 48–49
  resources 84
Inverclyde Group 7, 9–15, 83, 85
Inverleven Formation 58, 76
Ironhead Coal 30

Jedburgh Basalt/lava 17, 18, 22, 23, 24, 26
Johnstone 27, 29, 30, 34, 48, 51, 83–84
Johnstone Castle 52
Johnstone Clayband Ironstone 36, 45, 83
Johnstone Shell Bed 34, 37, 44, 45
Jordanhill Ironstone 83
Jubilee Shale Coal 34

kame 54, 64
Kelvin Formation 55, 82
Kelvin, River 88
Kelvin valley 55, 57–58, 60, 63–64, 66–70, 82, 84–85, 87–88, 91
Kelvinhead 64
kettleholes 54, 64
Kilbarchan 25, 48, 84
Killearn 3, 4, 57, 74, 86
Killearn Borehole 69, 70, 71, 73, 74, 75
Killearn Formation 58, 64, 71
Kilmannon 21
Kil,ammon Reservoir 22
Kilmaronock Formation 58, 79, 82
Kilpatrick Block 90, 93, 95
Kilpatrick Hills 10, 12–17, 27, 55, 57, 64, 76–77, 81, 85
Kilpatrick Hills succession 19–26
Kilsyth 57
Kilsyth Block and Hills 17, 24
Kilsyth Coking Coal 34, 83
Kilsyth Trough 50, 51
Kiltongue Coal 40, 83
Kiltongue Musselband 46
Kinderscoutian Stage 7
King Blackband Ironstone 37, 45
King Coal 44
Kinnesswood Formation 7, 8, 9–13, 83
Kippen 91
Kipperoch Borehole 12
Kirkwood Formation 7, 15, 27, 30
Knightswood Gas Coal 34, 36, 83, 86–87
Knightswood Under Coal 37
Knockshannoch 22
Knockupple 12, 21
Knockupple Vent 22
Knowehead Lavas 19, 22-3, 24
Knox Pulpit Formation 86, 92

Lady Ann Coal 30
Ladygrange Coal 40, 46
lake, glacial 66–69
  ice-dammed 59, 72
  shoreline 61, 74, 76
'Lake Clydesdale' 64, 66–69
'Lake Kelvin' 66–67, 69
landfill 87
Lands of Rylees Farm 57
landslip and solifluctued deposits 55,

76–78
Lang Craigs 21, 22, 76
Langbank 57, 66
Langsettian (Westphalian A) Stage 7
lavas 17–19, 21–26, 84
Law Formation 58, 82
Lawmuir Formation 7, 15, 27–31, 43, 47–49, 83–84
Lawmuir Borehole 27, 43
leaching 30
  containment of 87
lead minerals 14
Lenisulcata Chronozone 40
Lennox Castle 27, 29
Lennoxtown 63, 86, 90
Lenzie-Torphichen dyke 48, 93
Lesmahagow Inlier 92, 93
Leven valley 55
Levern Towers 74
Lewisian 90, 91, 95
Lillie's Sandstone 31
Lillie's Shale Coal 31, 83, 84
Lily Loch 21
Limecraigs see Nethercraigs
limestone
  dolomitic 29, 38
  Lower Carboniferous 6, 7, 8, 27, 29, 30, 31, 43–44, 45
  resources 83, 84, 85
  Upper Carboniferous 34, 38, 39, 43–46, 47
Limestone Coal Formation 6, 7, 34–37, 43, 44, 48, 49, 50, 51, 53, 83, 84, 85
linear vent system 17, 18, 50
Linn of Baldernock 29, 49, 83
Linn Well 23
Linwood 34, 48, 53, 54, 57, 78, 83
Linwood Basin 51
Linwood Borehole 62, 69, 70, 78, 79, 81, 82
Linwood Formation 58, 69, 70, 71, 72, 77–79
Linwood Moss 79, 81, 82, 85
Linwood Moss Wood Borehole 79
Linwood-Paisley embayment 57, 79
Linwood Shell Bed 37, 45
listric fault 92
lithology
  Limestone Coal Formation 34–37
  Lower Limestone Formation 31–33
  Upper Limestone Formation 37–39
lithostratigraphy
  Lower Carboniferous 6, 7
  Quaternary 58
  Upper Carboniferous 6, 7
Little Corrie 15, 57
Loch Humphrey Borehole 14, 15, 21, 22, 43
Loch Lomond Basin 78, 79
Loch Lomond Stadial (Readvance) 54–55, 67, 72–78
Lochend Cottage 4
Lochinch 52
lodgement till 61
Longhaugh Formation 58, 81
Longhaugh No. 20 Borehole 81

Lower Carboniferous (Dinantian)   6, 9–33, 48, 49, 50, 51, 88, 90, 91, 92
Lower Castlehead Coal   29, 30, 83
Lower Clyde valley   79–81
Lower Coal Measures   6, 7, 8, 29, 40, 41, 46, 51, 83
Lower Devonian   3, 16, 51, 83–84, 89–92, 95
Lower Drumgray Coal   46
Lower Dykebar Coal   29
Lower Garscadden Ironstone   36–37
Lower Househill Clayband Ironstone   31, 83
Lower Limestone Formation   6, 7, 31–33, 43, 47, 48, 49, 83, 85
Lower North Campsie Lavas   18, 19, 25
Lower Possil Ironstone   37, 45, 83
Lower South Campsie Lavas   18, 19, 25
Lowstone Marine Band   7, 8, 34, 39, 40
Lyoncross Limestone   37, 38, 39, 45, 83

magnetic investigations   88, 90, 91, 93, 94
Main Hosie Limestone   31
'Main Late-glacial Shoreline'   79
Mains of Kilmaronock Borehole   72, 73, 78, 79
malachite   24
man-made deposits   82
Mansionhouse Road   69
marble   41, 46
marine incursions/transgressions/deposition
  Lower Carboniferous   6, 8, 27, 30–31, 33
  Quaternary   54, 58, 61
    Loch Lomond Stadial   74, 76, 77–78, 79–81
    Windermere Interstadial   66, 67, 69–72
  Upper Carboniferous   34, 37, 38–39, 40–41, 42, 43, 44, 46
Markle Basalt/lava   17, 18, 19, 22, 23, 25
marl   30
Marsdenian Stage   7
Marshall Moor Lavas   25
Maryhill   39
Maryhill Borehole   48
Mavis valley   87
Meikle Caldon   72
meltwater deposits   64
Merryhead Coal   30
methane   86, 87
microfaulting   61
microfossils   46–47, 54, 63, 69
Mid Hosie Limestone   31, 44
mid-Dinantian unconformity   50, 51
Middle Coal Measures   6, 7, 40–42, 46, 83
Mill Coal   46
millet seed grains   4
Millfaid   4
millstone   12, 13
Milngavie   27, 29, 43, 48, 49, 84
Milngavie-Kilsyth Fault   49, 51, 53, 90
Milngavie Marine Band   31

Milngavie No. 2 Borehole   31
Milngavie No. 4 Borehole   58
Milngavie No. 5 Borehole   31, 49, 58
Milngavie No. 6 Borehole   49
Milton of Buchanan Water Borehole   85
Milton Hill   84
Minehead Coal   30
miospores and miospore zones   8, 31, 46, 47
Misty Law Trachytic Centre   25
Moor Rock Coal   39
moraine   64, 66, 72, 74
Moss Cottage   81
Moss Cottage Borehole   79
mudstone resources   84
Mugdock   49
Mugdock Dyke   48
Mugdock Lavas   21, 23, 24, 25
mugearite   17, 19, 21, 22, 23, 25, 26
Muir Toll Burn   18
Muirhouse   23, 27
Mull regional dyke swarm   49
Murroch Glen   43

Namurian   8, 34, 37–39, 40, 43–45, 51
native copper   24
natrolite   24
Necropolis Hill   39, 48, 49
Neilson Shell Bed   43, 44, 45
Neilston Block   90, 93
Nethercraigs (Limecraigs)   30, 31, 52, 83
New City Road   61
New Jordanhill Blackband Ironstone   37
Newlands   31, 83
Newton Coal   29, 30
Nine-Foot Coal   30
Nitshill   30
Nitshill Sandstone   34, 84
Nitshill No. 6 Marine Band   8
Noddsdale Volcaniclastic Member   25
nodular carbonate   14
normal fault   53
North Brae Coal   39
North Campsie Lavas   18, 24, 25
North Campsie Linear Vent System   17, 19
North Kilbowie Farm   57

Oakshaw Hill   54
oil shale   31
oil shale resources   84
Old Jordanhill Blackband Ironstone   36–37
Old Kilpatrick   15, 23, 25
Old Patrick Water   49
Orchard   45
Orchard Limestone   37, 38, 39, 44, 45, 83
Orchard Quarry   44
outwash deposits   60–61
overbank deposits   9, 14, 31
Overtoun Burn   10, 14
Overtoun Sandstone Member   14

Paisley   15, 23, 24, 27, 29, 30, 31, 34, 48, 51, 52, 53, 54, 57, 69, 70, 78, 79, 83, 84, 85
Paisley Formation   58, 67, 69, 70, 71, 72, 75
Paisley Moss   79
Paisley Ruck   24, 34, 51, 52, 53, 85, 90, 91, 95
palaeocurrent measurements   27
palaeogeography
  Coal Measures   40, 42
  Limestone Coal Formation   37
  Lower Carboniferous   30–31, 33
  Lower Limestone Formation   33
  Upper Limestone Formation   39
palaeontology
  Lower Carboniferous   8, 9, 38, 43
    Strathclyde Group   21, 27, 29, 30, 31
  Quaternary   48, 54
    Devensian   55–57
    Flandrian   78, 79, 81–82
    Loch Lomond Stadial   72, 74, 76, 77, 78
    pre-Dimlington   61, 63
    Windermere Interstadial   69, 70, 72
  Upper Carboniferous   6, 8, 43–47
    Clackmannan Group   34, 37, 38, 39
    Coal Measures   40, 41, 44-5, 46
palaeoslope   5, 26, 50
palaeowind directions   5
paralic deltaic conditions   37
Park Hill Vent   16–17, 24
Parks of Drumquhassle   72, 74
Partick   53
Passage Formation   6, 7, 8, 39–40, 46, 48, 51, 92
peat and peat moss   55, 58, 70, 76, 78, 79, 81–82, 85
pedogenic processes   9
Pendleian Stage   7, 34, 37, 43, 45
peneplanation   33
periglacial conditions   55, 61, 64, 72, 77
Permian   51, 90, 92–93, 95
Permo-Carboniferous intrusive rocks   90, 91, 92-3, 95
phonolytic trachyte   19
physical properties, investigations into   92–93
'pin-stripe' laminations   4
pisolitic structures   9
pit bing   82
Plean No. 1 Limestone   38, 39
Plean No. 2 Limestone   38, 39, 40
Plean No. 3 Limestone   38
Pliocene   57
plug   15, 17
pollen   79, 82
pollution   87
Polmadie   84
Portnauld   51, 78
Portnauld Farm railway cutting   78
Possil Ironstone   37, 45, 83
Possil Main Coal   34, 36, 83
Possil Wee Coal   34, 36
potassic alkali basalt series   26
Pow Burn   82

pre-Dimlington Stadial
    deposits   57–63
       glaciofluvial outwash deposits   60–61
       glaciolacustrine deposits   61, 63
    lithostratigraphy   58
Priesthill Fault   53
proglacial lake   66
Prospecthill Borehole   42, 46
pyroclastic deposits   15, 21

Quarrelton Thick Coal   30, 48, 52, 83
quartz-conglomerate   27, 29, 50, 84
quartz-dolerite intrusions   48, 51, 84, 90, 91, 92–93, 95
quasimarine conditions   30, 34, 37, 39, 50
Quaternary   54–82
   Devensian topography and palaeontology   55–57
   economic geology   84–85, 86
   geophysical investigations   90, 91
   Late Quaternary history   54–55
   lithostratigraphy   58
   sea level   66, 71, 76, 78, 79–81
Queenslie see Vanderbeckei

radiocarbon/radiometric dating of Quaternary deposits   61, 71, 74, 76–77, 78, 79, 81–82
raised mire, moss and peat   30, 55, 78, 85
raised shoreline   71–72
Ralston Hill   57
reclamation, land   87
reflection, seismic   91
refraction, seismic   91–92
refractory clay   84
regional tensional stress regimes   27
Renfrew   34, 48, 76, 79
Renfrew Road bridge   70
Renfrewshire Hills and succession   15, 17, 23, 25, 54, 64, 90
resistivity, seismic   92
rhinoceros, woolly   61
rhyolite   17, 19, 26
Richey Line   91
Rigangower   21
Rigangower Quarry   43
Riggin Anticline   91
ring-faulting   17
Robroyston   76
roches moutonnée   57
rock-cored drumlin   63
Roman Cement Limestone   39
Ross Formation   58, 64, 66, 67
Rutherglen   71
Rutherglen Fault   42

sabkha   14
St Vincent Street   67
saline lake   14
sand and gravel   55, 58, 84–86
Sandholes   30
sandstone resources   83–84, 86
Saucel Hill   30
Saughen Braes Lavas   17, 21, 22, 25, 26
scoria bombs   15

scoriaceous agglomerate   18
Scotstoun   76
Scotstoun House   71
scree   77
sea levels in Quaternary   66, 71, 76, 78, 79–81
sea-bottom mud   70
Second Hosie Limestone   31
segregation veins   49
seismic investigation   91–92
septarian cracks   14
Sergeantlaw Lavas   23–24
shallow mine-workings   86
sheet-flood deposits   14
shell bed   34, 37, 43, 44, 45, 54, 58–59
Shettlestone Sandstone   42
Shieldhall   66, 67, 68, 69, 70, 81
Shiels   76
Shilton March Burn   78
sideritic laminae   34
Skipsey's see Aegiranum
slag tip   82
Slatehouse   3
Smiddyhead Coal   30
sodic alkali basalt series   26
Soho Street   40
solifluxed deposits   76–80
South Arthurlie Coal   39
South Brae of Campsie   29, 83, 84
South Campsie Lavas   18, 25
South Campsie Linear Vent System   17, 18, 19
spatter   15
sphaerosiderite   29, 30
Spout of Ballagan   15
Springburn   39, 76–77
Springburn Brickworks   84
springs   86
Station Wood, Killearn   74
Stobcross railway cutting   67, 69
Stockiemuir   13
Stockiemuir Sandstone Formation   4–5, 86, 92
stoop and room workings   29, 86
Strath Blane   10, 13, 14, 15, 17, 24, 66, 71, 72, 74, 78–79, 82, 84
Strath Blane Basin   78–9
economic geology 84
Strathblane   22, 84
Strathblane-Balmore Tunnel No. 8 Borehole   27
Strathclyde Group 7, 15–31, 86
Stratheden Group   4–5, 11
Stockiemuir Sandstone Formation   4–5, 86, 92
Strathgryfe Lavas   25
Strathmore Group   3
Strathmore Syncline   3, 50, 51, 93
striae, glacial   54, 57, 63
structure   1, 50–53
subglacial stream   57
Summerston   87
Summerston Fault   53

Tambowie Lavas   21, 23, 25
Teith Formation   3
tensional stress regimes   27

tephra   17, 21, 22, 24, 25, 26
terminal moraine   66, 72
Tertiary intrusive rocks   49
thompsonite   24
Thornliebank   36, 44
till   54, 57, 58, 59, 60, 61, 63, 64, 66, 71, 72, 74, 76, 77, 78
Titwood No. 2 Pit   31
Tomibeg   76
Top Hosie Limestone   6, 7, 8, 31, 43–44, 45
Torrance   57, 66
Tournaisian series   7
Townhead   72
trachybasalt   17, 18, 19, 22
trachyte   26
tuff   15, 18, 19, 21
tundra conditions   77
tunnel, glacial   64
Twechar Dirty Coal   34
Twechar Upper Coal   34
Tweeniehills railway cutting   78

unconformity, mid-Dinantian   50, 51
Upper Campsie Lavas   18, 25
Upper Carboniferous (Silesian)   7, 8, 34–42, 49, 51, 83, 86, 90
Upper Castlehead Coal   29
Upper Coal Measures   6, 7, 29, 41, 42, 46, 51, 52, 53, 83, 84
Upper Devonian   3–6, 83–84, 86, 90, 92
Upper Drumgray Coal   40, 46
Upper Drumgray Musselband   46
Upper Dumbreck Sandstone   38, 39
Upper Dykebar Coal   29
Upper Garscadden Ironstone   36–37
Upper Hirst Coal   39
Upper Househill Clayband Ironstone   31, 83
Upper Limestone Formation   6, 7, 37–39, 45–46, 47, 48, 49, 51, 83, 84, 85
Upper North Campsie Lavas   18, 19, 25
Upper Orchard Limestone   38
Upper Possil Ironstone   37, 83
Upper South Campsie Lavas   18, 19, 25

valley glacier   66
Vanderbeckei (Queenslie) Marine Band   6, 7, 40, 41, 42, 46
varves   61, 69, 72, 74, 79
Victoria Park   45
Victoria Pit   83, 84
Virgin Upper Coal   41
Viséan Series   7
vivianite   79
volcanic activity, centres   6, 15
volcanic detritus   19, 27
volcanic vents
   Clyde Plateau Volcanic Formation   15, 16–17, 18, 24, 84
   Kilpatrick Hills   19, 20, 21, 22, 23
volcaniclastic material   15, 21, 22, 23–24, 29, 30, 43, 47

Wallace Pit, Elderslie   30

waste disposal sites 87
Waterhead Central Volcanic Complex 17, 19, 21, 25, 26
Waterside high 91
Waulkmill Glen 38, 83
Western Campsie Fells succession 17–19, 22-3, 24, 25
Westmarch and Shortroods Clayfield 57
Westmuir Sandstone 42
Whangie, The 77
White Cart Water 48, 57, 82, 84
White Limestone *see under* Baldernock Limestone
Whittliemuir Midton Loch 23
Wilderness Gravel Pit 60–61, 85
Wilderness Landfill 87
Wilderness Plantation 63
Wilderness Till Formation 58, 59, 60, 61, 63, 66, 69, 71, 84
Williamwood 64, 84
Willowbank Crescent Borehole 34
Windermere Interstadial (deglaciation) 54, 64–72
  glacial meltwater deposition 64
  ice wastage deposits 64
  lithostratigraphy 58
  marine inundation 69–72
Windmillcroft Dock 67–68, 81
Windyhill 78, 83

Yeadonian Stage 7
Yoker 48

zeolite/zeolitisation 17, 21

BRITISH GEOLOGICAL SURVEY

Keyworth, Nottingham NG12 5GG
0115 936 3100

Murchison House, West Mains Road, Edinburgh
EH9 3LA   0131-667 1000

London Information Office, Natural History Museum
Earth Galleries, Exhibition Road, London SW7 2DE
0171-589 4090

The full range of Survey publications is available through the Sales Desks at Keyworth and at Murchison House, Edinburgh, and in the BGS London Information Office in the Natural History Museum (Earth Galleries). The adjacent bookshop stocks the more popular books for sale over the counter. Most BGS books and reports can be bought from The Stationery Office and through Stationery Office agents and retailers. Maps are listed in the BGS Map Catalogue, and can be bought together with books and reports through BGS-approved stockists and agents as well as direct from BGS.

The British Geological Survey carries out the geological survey of Great Britain and Northern Ireland (the latter as an agency service for the government of Northern Ireland), and of the surrounding continental shelf, as well as its basic research projects. It also undertakes programmes of British technical aid in geology in developing countries as arranged by the Department for International Development and other agencies.

The British Geological Survey is a component body of the Natural Environment Research Council.

Published by The Stationery Office and available from:

The Publications Centre
(mail, telephone and fax orders only)
PO Box 276, London SW8 5DT
General enquiries 0171 873 0011
Telephone orders 0171 873 9090
Fax orders 0171 873 8200

The Stationery Office Bookshops
59–60 Holborn Viaduct, London EC1A 2FD
temporary until mid 1998
(counter service and fax orders only)
Fax 0171 831 1326
68–69 Bull Street, Birmingham B4 6AD
0121 236 9696   Fax 0121 236 9699
33 Wine Street, Bristol BS1 2BQ
0117 9264306   Fax 0117 9294515
9–21 Princess Street, Manchester M60 8AS
0161 834 7201   Fax 0161 833 0634
16 Arthur Street, Belfast BT1 4GD
01232 238451   Fax 01232 235401
The Stationery Office Oriel Bookshop
The Friary, Cardiff CF1 4AA
01222 395548   Fax 01222 384347
71 Lothian Road, Edinburgh EH3 9AZ
(counter service only)

Customers in Scotland may
mail, telephone or fax their orders to:
Scottish Publications Sales
South Gyle Crescent, Edinburgh EH12 9EB
0131 228 4181   Fax 0131 622 7017

The Stationery Office's Accredited Agents
(see Yellow Pages)

and through good booksellers